T0205510

Textile Science and Clothing Technology

Series editor

Subramanian Senthilkannan Muthu, Hong Kong, Hong Kong SAR

More information about this series at http://www.springer.com/series/13111

Rajiv Padhye · Rajkishore Nayak
Editors

Acoustic Textiles

Editors
Rajiv Padhye
RMIT University
Melbourne, VIC
Australia

Rajkishore Nayak
RMIT University
Melbourne, VIC
Australia

ISSN 2197-9863 ISSN 2197-9871 (electronic)
Textile Science and Clothing Technology
ISBN 978-981-10-9364-7 ISBN 978-981-10-1476-5 (eBook)
DOI 10.1007/978-981-10-1476-5

Printed on acid-free paper

This Springer imprint is published by Springer Nature
The registered company is Springer Science+Business Media Singapore Pte Ltd.

Preface

This book was motivated by the inspiration gained during our literature survey in acoustic textiles. A prominent gap was felt to gather concise information in one place. The information available were scattered in some published papers and conference proceedings. Although some books covered the fundamentals of sound and techniques of noise control, there were a few literatures especially a textbook that provides knowledge on the topic as a whole. The editors felt that there is a necessity to include the relevant information in the form of a book, which can provide in-depth knowledge to the readers.

The editors' main aim is to provide in-depth knowledge to all the personnel involved in the production and installation of acoustic textiles. Acoustic textile is multidisciplinary, which needs the knowledge of sound as in physics, engineering principles of noise control, understanding the fundamentals of textile materials, civil engineering and interior designing. As the application of fibrous materials in noise control is increasing in recent years, the knowledge of textile technology is essential for effective design of acoustic materials.

We have tried to make this book informative by covering all the aspects of sound, from its generation, propagation to methods of noise control; from the impact of noise on health to regulations governing the noise control. Various materials used including the manufacturing methods and their evaluation techniques; and influencing factors are highlighted. The major applications of acoustic textiles in transportation and civil engineering are also covered. Standards related to acoustic materials and noise regulations are also covered in this book. It also includes the organizations related to acoustic textiles, list of journals and recommended conferences around the globe.

This book provides state-of-the-art knowledge on fundamentals of sound, principles and materials used for acoustic applications. Contributors of this book are subject experts in their respective fields. The authors have shared their knowledge and experience through their latest research, which will serve as a single source of knowledge in the field. The book is written in simple language so that the people with minimal knowledge can easily understand the subject matter. It is anticipated that the book with some scientific knowledge and fundamental principles will dissipate knowledge to the acoustic engineers, students and research community.

The editors extend their sincere thanks to all the experts who have contributed valuable time and effort for the successful completion of the chapters, in order to create this important body of work. The editors are also thankful to Springer Publishing for their support for making this book a reality. We hope this book will help the academicians and researchers in the field of acoustic textiles in acquiring information on the subject matter.

Melbourne, Australia

Rajiv Padhye
Rajkishore Nayak

Contents

Acoustic Textiles: An Introduction

Rajkishore Nayak and Rajiv Padhye

Abstract Noise pollution can be defined as the excessive sound that can imbalance the activity level of human or animal life. Excessive noise has adverse effect on the physiological and psychological effects on human beings that can cause permanent hearing damage, increased stress, reduced efficiency at work, disturbance of the sleep pattern and interference in the communication. Noise is considered to have negative effect in a wide range of modern engineering areas. The problem of noise pollution becomes more aggravated in an enclosed environment. The competitiveness of products such as cars and washing machines is being considered by the noise level as one of the factors. This chapter describes the global usage of acoustic textiles and the reasons for increased usage. The rules and regulations in different countries of the world are also covered in this chapter. The methods used for noise control and various materials used for the same are briefly covered. In addition, a list of standards relating to acoustics and governing bodies around the world dealing with acoustics is also discussed.

Keywords Acoustic textiles · Noise control · Porous materials · Acoustic standards · Recycling

1 Introduction

As described in the Oxford dictionary, the meaning of acoustic is "*Relating to sound or the sense of hearing*". Study of acoustics involves the section of science that focuses on all mechanical waves in gases, liquids and solids. The word "*acoustic*" is derived from the Greek word *ἀκουστικός* (akoustikos) that means hearing or ready to hear [1]. The synonym for acoustic in Latin is "*sonic*", on the basis of which the term sonics is used as a synonym for acoustics. As per ANSI/ASA S1.1-2013, acoustic is defined as: (a) Science of sound, including its

R. Nayak (✉) · R. Padhye
School of Fashion and Textiles, RMIT University, Brunswick, Melbourne, VIC, Australia
e-mail: rajkishore.nayak@rmit.edu.au

R. Padhye and R. Nayak (eds.), *Acoustic Textiles*, Textile Science and Clothing Technology, DOI 10.1007/978-981-10-1476-5_1

production, transmission and effects (biological and psychological effects); and (b) Those qualities of a room that, together, determine its character with respect to auditory effects [2].

Sound is an inherent component of our environment, from which we cannot escape. Even in the dark and deep ocean, creatures communicate using the water medium where the sound propagates over a long range. The existence of sound can be considered as negligible in places such as high vacuums, where there are no atoms to interact with each other; and in cosmic space.

In addition to the animals, sound is also essential for humans to communicate. Sound can be classified as pleasant or noisy. The pleasant form of sound is the music, which has emotional impact on human beings. On the other hand, the sound from a moving train or construction site is disliked by everyone. This is because the sound energy from these sources exceeds the threshold of pleasantness and becomes a noise. Noise is the unwanted loud sound that eliminates the wanted sound and disturbs people and animals in many ways.

Excessive noise has adverse effect on the physiological and psychological effects on human beings that can cause: permanent hearing damage, increased stress, reduced efficiency at work, disturbance of the sleep pattern and interference in the communication [3–5]. Noise is considered to have negative effect in a wide range of modern engineering areas. The problem of noise pollution becomes more aggravated in an enclosed environment. The competitiveness of products such as cars and washing machines is being considered by the noise level as one of the factors [6].

Noise pollution can be defined as the excessive sound that can imbalance the activity level of human or animal life. The noise pollution can be classified as outdoor and indoor. The outdoor noise pollution is caused in the environment by transportation systems, machines and construction, whereas indoor noise can be caused by indoor activities, concerts or running of motorized equipment. Both outdoor and indoor noise pollutions are harmful to human beings as well as animals. Poor urban planning may lead to both outdoor and indoor noise pollution.

From these above discussions, it is imperative that the noise should be effectively controlled. These problems can be overcome by fabricating quite machines, designing buildings and structures with acoustic absorbing materials and applying noise control measures. Furthermore, strict regulations and standards can be used for effective noise control in domestic, commercial and other public places. This chapter describes the reasons for increased usage of acoustic textiles and their global usage. The rules and regulations in different countries, the methods used for noise control and various materials used for the same are briefly covered. Furthermore, a list of standards relating to acoustics and organizations around the world dealing with acoustics are also discussed.

2 Global Usage of Acoustic Textiles

There are several application areas of acoustic materials in building and construction; transportation; and industrial. The global consumption of acoustic textiles has been rapidly increasing due to various reasons, which are discussed in the following section.

2.1 *Reasons for Increased Usage*

The consumption of acoustic textiles around the world is rapidly increasing due to increased application areas, technological advancements, growing demand and stricter regulations in many countries on noise regulation. One of the major driving sectors for acoustic textiles is the transportation. Furthermore, the increasing application in residential and commercial buildings has accentuated the global usage. The increasing demand for acoustic textiles in developing countries in addition to the growing preference of manufacturers for acoustic insulation products in transportation, are the major areas of acoustic applications. In future, the industrial applications of acoustic materials in areas such as oil, gas and power generation have a wide scope for its expansion.

Population growth, stricter regulations in the construction of new buildings, increased environmental concern and sustainability are also the driving forces for increasing acoustic applications. Materials such as porous textile structures, stone wool, glass wool and foamed plastic are expected to be in high demand for various acoustic insulation applications. In 2013, stone wool was the most consumed acoustic insulation material. Factors such as excellent thermal insulation, superior fire resistance, better sound insulation and ease of installation contributed to this.

2.2 *Statistical Data*

Between 2014 and 2019, the market for acoustic insulation is expected to witness a compound annual growth rate (CAGR) of 5.8 %. This growth will generate a revenue of $4159.04 million (USD) by 2019 only in the areas of industrial, transportation, building and construction. The global value of acoustic materials in building and construction, the largest area of acoustic insulation across the world, is projected to reach $1648.43 million by 2019.

In 2013, the European region was the largest market for acoustic materials that constituted 37.88 % by value of the total global acoustic insulation material consumption. Asia-pacific region is the other potential market due to increased manufacturing base in this region. In Asia-pacific region, transportation, building and construction sector are rapidly growing, which is the major reason for growing demand for acoustic insulation materials in the region. As the demand for comfort and peaceful life is growing day by day, the demand for acoustic insulation materials is also increasing to cater the demand.

Some of the global market players for manufacturing and supply of acoustic materials include Rockwool International A/S (Denmark), Knauf Insulation (Germany), Carpet Concept (Germany), Vic Saint-Gobain (US), Owens Corning Corporation (US), Johns Manville Corporation (US), Armacell (Germany), BASF SE (Germany), Vicoustic (Portugal), Arper (Italy), Texaa (UK), Tela-design (Germany), Acoustic Pearls (Germany), Gotessons (Sweden), Spigo GROUP (France), Valoa by Aurora (Finland), Paroc (Finland) and Fletcher Insulation (Australia).

3 Noise Generation and Its Impact

The problem of noise pollution is increasing day by day around the world. Increased application of different mechanical and electrical equipment in domestic and industrial sector has led to higher noise level. High demands on newly constructed buildings due to urbanization and industrialization have accentuated the problem of noise pollution. The places, which need to be quiet such as hospitals, homes, seminar rooms, meditation places and more importantly schools are affected by noise pollution in many places. This is more alarming in developing countries that are lacking strict legislations. Once considered as harmless, noise pollution is emerging as a great threat in the modern world to people's health as it can lead to mental illness, divert the attention and lead to physical ailment.

3.1 Sources of Noise

Noise is mainly produced by many mechanical systems including vehicles, construction work, large buildings, home appliances, rail transport, airplane and industrial machineries [7–9]. In addition, the leisure sources, sports events and shooting ranges also cause noise pollution [10–12]. In many countries, one of the major sources of noise is traffic, which is generated from the engine, friction of tyre with road surface, aerodynamic and braking elements [13–15]. Traffic noise is a major contributor to the noise pollution in developed and developing countries [16]. The traffic noise has increased manifold compared to what was in the previous century, although there are hybrid vehicles. This is because of increase in the number of vehicles, higher speed limits and increased number of roads and freeways. The noise pollution by aircrafts is also a major problem in many countries. The aircraft noise is produced during different phases of its flight such as take-off, landing, while taxing, in its path or even when parked in the ground.

The aircraft noise can be classified as aerodynamic noise, engine noise and noise from aircraft systems. The aerodynamic noise is generated by the movement of the air around the airplane control surfaces and the fuselage. This noise level depends on the speed and altitude of the aircraft. This can be substantially higher for a low-flying military aircraft at high speed. The jet engines produce majority of the noise during take-off, climb and landing. Similarly, the noise from propeller aircrafts such as helicopters is aerodynamically generated from the main and tail rotors or mechanically generated by the main gearbox and the transmission chains. The noise from aircraft systems include the noise from the conditioning systems, cabin and cockpit pressurization. The auxiliary power unit used to start the engine or to provide power while the aircraft is on the ground is the other source.

Industrial noise generated from various machineries and processes is another source of noise pollution [17]. Various industries such as textile, metal industry (iron and steel), saw mills, foundries and crushing mills are the major contributors of noise pollution. In these industries, moving components such as vibrating panels, rotors, gears, fans, generators and engines generate high level of noise. The mechanism of noise generation by these components depends on the equipment and operation such as riveting, crushing, blasting (mines and quarries), punch presses, foundries, lathe, drilling, pneumatic equipment, drop forges, plasma jets, sand blasting, tumbling barrels, boiler making, grinders, milling machines and furnaces.

Household noise and social events can also be major noise sources in many countries. Household items such as washing machine, air conditioners, vacuum cleaners, grinders and music systems including television are the source of noise. Social events, parties, worships, concerts in open areas and clubs can also produce substantial amount of noise depending on the society and country, which affect the nearby residents. There is also significant contribution by the entertainment venues as amplified music and high volume sounds create noise pollution for people in the proximity.

3.2 *Impact of Noise Pollution*

The daily activities of many people at work, school, home and during leisure time are disturbed by excessive noise. In Europe, studies have revealed that every year, fifty thousand deaths and about a quarter of a million cases of cardiovascular diseases are reported relating to traffic noise [18, 19]. In Denmark, it has been found that 5 % of all the stroke cases reported are caused by traffic noise [20]. According to the World Health Organization (WHO), the effect of noise pollution on health and environment is second next to air pollution.

There are several adverse effects of regular exposure to elevated noise level to humans as well as to animals. High levels of noise can lead to hypertension, hearing loss, sleep disturbances, heart diseases, increased stress and hinders performance at work and children's learning [21–23]. The effect of noise on humans can be classified into three groups: (a) physical effect, (b) physiological effect and (c) psychological effect. The physical effects can directly influence the health of a person, and adverse effects such as permanent hearing loss can happen. The physiological effect can also adversely influence the health, and problems such as high blood pressure and increased stress level may be the output. The psychological effects influence a person's welfare that can lead to annoyance, distraction and complaint. In order to avoid these, many people try to move to quieter places paying any cost.

Hearing loss caused by noise pollution is permanent, which can be completely prevented. In many countries, the lack of proper regulations or non-implementation of the regulations leads to the hearing loss of millions of workers. Strict legislations can help in preventing the noise-induced hearing loss. One of the regulations by the National Institute for Occupational Safety and Health specifies a maximum exposure of 85 dB per eight hours shift per day per worker.

The aircraft noise can significantly affect human health as established by several studies [24, 25]. An average sound level of ≥40 dB from the aircrafts can lead to minor health impairment. A daytime sound in excess of 60 dB or a night time sound of 55 dB can lead to the risk of heart attack. In addition to the people living to the aircraft path, people inside the aircraft also suffer noise pollution. The noise levels for the people inside the plane (e.g. Airbus A321) can be a minimum of 65 dB while taxing and 80 dB during the journey. The former level is considered as 20 dB higher than the acceptable maximum for an office, and the latter as 20 dB lower than the industrial specifications.

Industrial noise pollution can lead to hearing impairment and even occupational hearing loss in addition to the other adverse health effects such as cardiovascular diseases, annoyance, increased accident in work place, anti-social behaviour and increased aggression. It can also cause hypertension, tinnitus and vasoconstriction. Prolonged exposure to a noise level of 85 dB or higher can cause increased blood pressure, elevated stress level and lead to permanent hearing loss. The greatest noise level that causes no detectable adverse alteration of morphology, functional capacity, growth, development or lifespan of the target organism is known as the "no observed adverse effect level (NOAEL)".

It is important to define threshold levels of noise in order to evaluate the health consequences. Thresholds such as $L_{\mathrm{Amax,\ inside}}$, $L_{\mathrm{Amax,\ outside}}$ and $L_{night,\ outside}$ are defined and the values are specified in Tables 1 and 2 [26]. Various health impacts of noise are described in Fig. 1 [26].

In addition to the humans, noise pollution can cause harm to the animal life by interfering with their navigation system, reproduction system and can lead to permanent hearing loss [27]. In some instances, it can even lead to the risk of death by altering predator and prey detection and avoidance. The noise pollution can lead to the reduction of usable habitat, which may lead to the extinction of the endangered species. The loud sound of military sonar has resulted in the death of certain species of whales [28]. With the presence of unexpected noise, different animal species may need to communicate loudly, which is termed as Lombard vocal response. If the sound is not loud enough, it is being masked by the noise. These sounds not received by the other animals of the same species may face a danger to them or loss some constructive signal. If one animal species communicate loudly, the other animals also need to be louder to communicate, resulting in ecological imbalance.

Table 1 Various effects of noise [26]

Effect		Indicator	Threshold, dB
Biological effects	Change in cardiovascular activity	[a]	[a]
	Electroencephalogram (EEG) awakening	$L_{Amax, inside}$	35
	Motility, onset of motility	$L_{Amax, inside}$	32
	Changes in duration of various stages of sleep, in sleep structure and fragmentation of sleep	$L_{Amax, inside}$	35
Sleep quality	Waking up in the night and/or too early in the morning	$L_{Amax, inside}$	42
	Prolongation of the sleep inception period, difficulty getting to sleep	[a]	[a]
	Sleep fragmentation, reduced sleeping time	[a]	[a]
	Increased average motility when sleeping	$L_{Amax, outside}$	42
Well-being	Self-reported sleep disturbance	$L_{Amax, outside}$	42
	Use of somnifacient drugs and sedatives	$L_{Amax, outside}$	40
Medical conditions	Environmental insomnia[b]	$L_{Amax, outside}$	42

[a]Although the effect has been shown to occur or a plausible biological pathway could be constructed, indicators or threshold levels could not be determined

[b]Note that "environmental insomnia" is the result of diagnosis by a medical professional whilst "self-reported sleep disturbance" is essentially the same, but reported in the context of a social survey. Number of questions and exact wording may differ

L_{Amax}: maximum noise level per event (Instantaneous effects of noise such as sleep disturbance due to passage of a lorry, aeroplane or train etc.) which is better understood with L_{Amax}

4 Noise Control

Noise can be controlled either by suppressing the noise creation or using sound absorbing materials that can reduce the sound energy. The effectiveness of sound absorbing materials can be improved by the use of barriers and internal enclosures. In the modern age of advanced computing, modelling and simulation can be used to investigate the effect of noise level on building structures and effective noise control. Strategies such as window glazing improvement, roof upgradation, fire place baffling and caulking construction can be evaluated by modelling and simulation.

4.1 Need for Noise Control

World Health Organization (WHO) specifies noise pollution as a global problem affecting the health of several people [29]. As discussed above, noise pollution can

Table 2 Effects of different levels of night noise on the population's health [26]

Average night noise level over a year, $L_{night,\ outside}$	Health effects observed in the population
Up to 30 dB	Although individual sensitivities and circumstances may differ, it appears that up to this level no substantial biological effects are observed. $L_{night,\ outside}$ of 30 dB is equivalent to the no observed effect level (NOEL) for night noise
30–40 dB	A number of effects on sleep are observed from this range: body movements, awakening, self-reported sleep disturbance, arousals. The intensity of the effect depends on the nature of the source and the number of events. Vulnerable groups (for example children, the chronically ill and the elderly) are more susceptible. However, even in the worst cases the effects seem modest. $L_{night,\ outside}$ of 40 dB is equivalent to the lowest observed adverse effect level (LOAEL) for night noise
40–55 dB	Adverse health effects are observed among the exposed population. Many people have to adapt their lives to cope with the noise at night. Vulnerable groups are more severely affected
Above 55 dB	The situation is considered increasingly dangerous for public health. Adverse health effects occur frequently, a sizeable proportion of the population is highly annoyed and sleep-disturbed. There is evidence that the risk of cardiovascular disease increases

$L_{night,\ outside}$: yearly average of night noise level outside at the façade (long-term effects of noise on health is understood by summarizing the acoustic situation over a long time period such as $L_{night,\ outside}$)

lead to various problems such as loss of hearing, reduced efficiency, tiredness, adverse social behaviour and even cardiovascular and psychophysiologic problems (Fig. 2). It is essential to have proper surrounding for the people to live in homes and the workers to work effectively in industries. Hence, the noise pollution needs to be reduced or eliminated to avoid the problems faced by millions of people.

Acoustic engineering deals with the design, analysis and control of sound and vibration. The main goal is to control unwanted noise or noise pollution. Various approaches such as the use of sound barriers, sound absorbers, silencers and buffer zones; redesigning of sound sources; acoustically engineering building and roadways and using hearing protection aids such as earplugs or earmuffs can help in controlling noise pollution. In addition to the noise control, acoustic engineering also covers the positive application of sound such as ultrasound in medicine and designing of digital sound synthesizers.

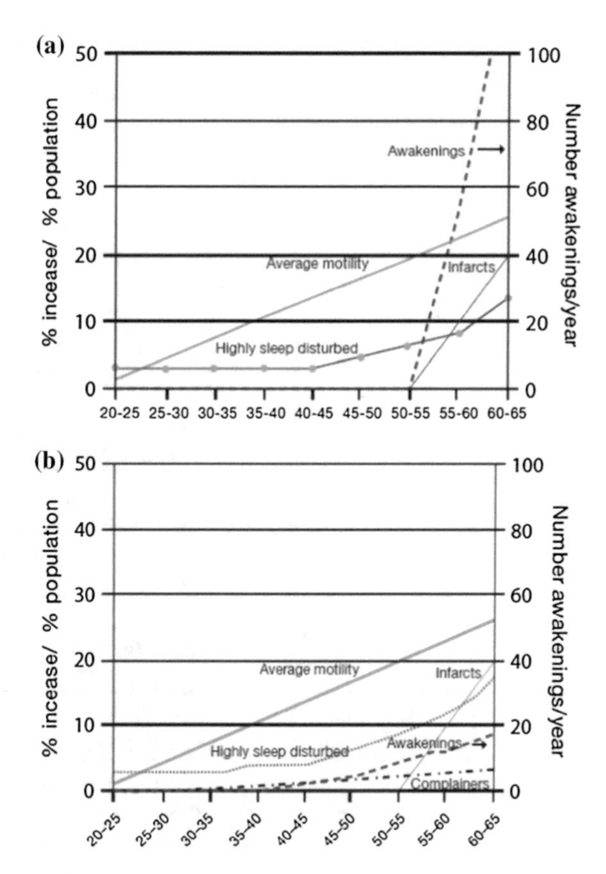

Fig. 1 Effect of night noise on human beings: **a** traffic noise and **b** aircraft noise [26]

4.2 Legality

The allowable noise level varies in different countries of the world depending on the community, social and cultural life of people. There are various regulations in many countries on the creation, regulation and control of noise. These noise regulations are established by national, state and even municipal levels of government that provide guidelines on the sound transmission. The noise regulation specifies the source, duration, amount of allowable noise and time restrictions of the day. The following section describes the regulations in the United States, Europe, Asia, Australia and New Zealand.

4.2.1 United States (US)

The Environmental Protection Agency (EPA) in the US had showed that about 44 million American population live in homes impacted by highway or aircraft noise, and 30 million American population is exposed to non-occupational noise high

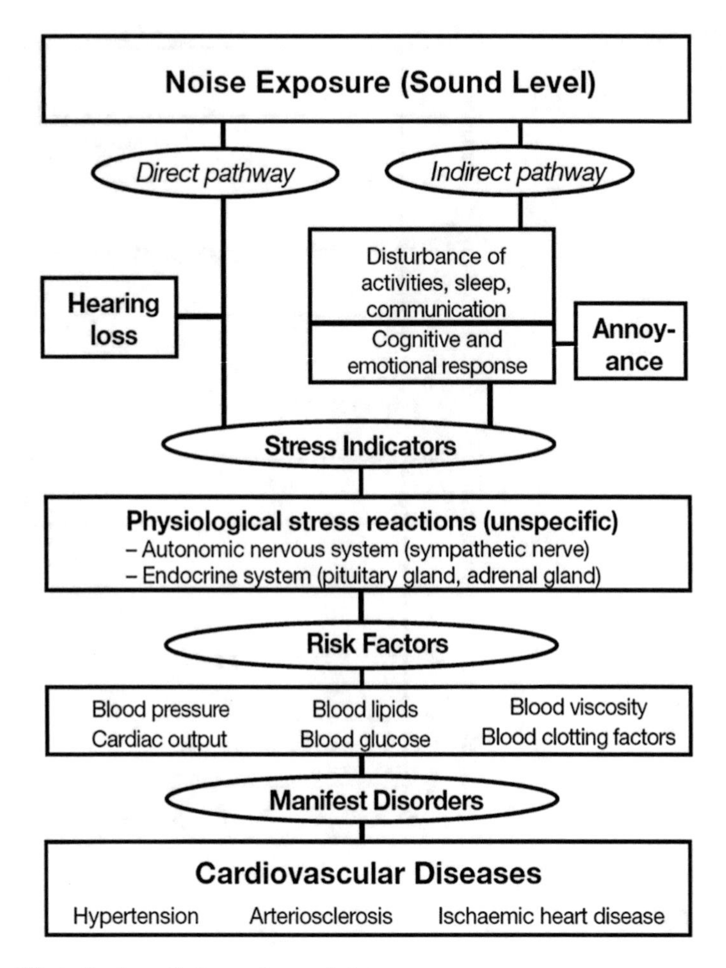

Fig. 2 Effect of noise pollution on human being

enough to cause hearing loss [30–33]. Hence, in order to control this, the National Environmental Policy Act (NEPA) was passed in 1969 and the Noise Control Act (NCA) was passed in 1972. Trespassing of sound above certain threshold intensity in residential areas during 9 pm to 7 am is prohibited in many cities as per the ordinance. Similarly, the law is also applicable for daytime, but with a higher threshold. The threshold may vary depending on the city and the state.

There are various federal standards that exist today in the US for highway and aircraft noise. In addition, there are stringent legislations relating to building codes, development of roadways and urban planning. The permissible noise levels relating to these items may slightly differ from council to council. The investigation and analysis of noise are controlled by the EPA. They can advice the public about the adverse effect of specific noise and, reply to enquires related to noise pollution. They can also evaluate the effectiveness of existing regulations in noise control.

4.2.2 Europe

Many countries in the Europe tried to establish laws equivalent to the US national law within the last three decades of twentieth century. For example, the noise legislations were passed in the Netherlands, France, Spain and Denmark in the years 1979, 1985, 1993 and 1994, respectively. However, there are many places in the US, where the legislations are legging compared to the Europe in spite of the regulations established earlier. In the European directive 2002/49/EC, article 10.1, the environmental noise has been defined, which is applicable to the European Union (EU). This directive provides guidelines to avoid, reduce or prevent the adverse effects of environmental noise. This directive specifies three actions: (a) to determine the environmental noise level, (b) to ensure that the information on environmental noise and their influence is provided to the public and (c) to reduce or prevent the environmental noise when necessary and preserve environmental noise when the quality is good. The Noise Regulatory Committee, European Environment Agency and the Noise Expert Group support the European Commission for implementation of the directive.

The directive is applicable to the noises that humans are exposed to in the parks, residential places and other quiet areas such as hospitals, schools and libraries. It does not include the noise caused by the exposed person himself, noise created by neighbours, noise from domestic activities, noise at work place and noise inside the means of transport or military activities in specific areas.

In the directive, it has been mentioned that in every five years the member states should prepare and publish noise maps and the action plan for noise control. While noise management action plans are being prepared, the member states' authorities are required to consult the concerned public. The action plan is meant for:

- agglomerations with more than 100,000 inhabitants
- major roads (more than 3 million vehicles a year)
- major railways (more than 30,000 trains a year)
- major airports (more than 50,000 movements a year, including small aircrafts and helicopters)

The main task of noise management will be implemented in 3-steps with an integrated approach. The first step will focus on the creation of noise maps for major roads, airports, agglomerations and railways by the competent authorities in the European member states. The second step will emphasise informing and consulting public about the concerns of the noise. The third and final step will focus on the action plans that will reduce the noise level.

Reducing traffic noise levels just by 3 decibels is similar to reducing the level of traffic by 50 %. This can help vast majority of the European citizens living in towns and cities or near major roads to achieve a major improvement in the health and quality of life. A 3 decibel reduction is quickly and easily achievable through technologies readily available for briefing New EU vehicle noise limits to car and truck manufacturers. Nearly one in every four cars and one in every three of the light trucks tested over the past 5 years already met the strictest standards proposed by the Commission.

Night sleep disturbances caused by excessive noise is a problem in many countries. The guidelines by WHO for night noise was published in 2009, which describes the allowable noise limits. As per the guidelines, the annual average sound in a residential area should be below 40 dB. People exposed to higher levels can suffer from mild health effects and insomnia. Longer exposure can lead to higher blood pressure or even heart attack. By recommending that countries introduce limits on night noise, the guidelines complement the 2002 European Union Directive on Environmental Noise, which requires countries to map hotspots and reduce exposure, but does not set limit values. The European Commission (EC) was a key partner and funded the project to develop the guidelines through its Directorate General for Health and Consumers.

4.2.3 Asia

Noise pollution is one of the problems faced in many Asian countries. Although there are several laws existing in controlling the noise in many countries, they are not properly implemented. Countries like Russia, China, Japan and India are the leading role players, but they are lagging much in this aspect compared to the US and Europe. Educational campaign and public awareness can help in mitigating this problem.

The method of acoustic control varies in many countries depending on the economy, culture and policies. Many countries are enforcing to use quieter machines and equipment, produce automotives with reduced noise level and follow the noise regulations prevalent in that area. For example, the **Buy Quiet** program that has been introduced in the US with an effort to mitigate occupational noise exposure, is being followed in many Asian countries. This program suggests individuals and industrials to buy machinery and power equipment with the lowest noise level. This program also encourages manufacturers to design quieter equipment. The companies that adopt this program are trying to create a working place with no harmful noise. Several companies are setting standard procedures with reduced noise level or automizing the equipment to reduce occupational health injury caused by noise pollution.

4.2.4 Australia/New Zealand

The Australia/New Zealand (ANZ) regulations for noise limits follow the EU noise directive in multiple areas. In April 2008, the Environment Protection and Heritage Council (EPHC) was agreed to pass a scheme for noise control by forming a working group. It was decided to adopt a staged approach to evaluate the scheme by implementing the scheme in a small range of articles. If this is found successful, it can be employed in other articles. EPHC approved the noise labelling scheme for fixed (domestic air conditioners and heat pump water heaters) and portable articles (leaf collectors, leaf blowers, lawn mowers, lawn edger, hedge trimmers and chainsaws) in November 2008. The labelling will help to reduce the noise if the noisy articles are used outdoors in residential areas. The council will set allowable noise limits for various other articles in different phases. This can be successful if strict legislations are set by the government.

The Environment Protection and Heritage Standing Committee (EPHSC) suggested in September 2009 that the ANZ noise labelling should be based on the EU noise directive and the heat pump water heaters and lawn edgers should be removed from the stage one as they are not included in the EU directive. EPHSC also agreed to employ an acoustical consultant to verify the accuracy of the measured noise levels for air conditioners, which are not covered in the EU scheme.

In ANZ, standards such as: AS 3781-1990 (Acoustics—Noise labelling of machinery and equipment) which is similar to ISO 4871-1984 (Acoustics—Noise labelling of machinery and equipment) was developed for stationery machinery and equipment. In Sect. 2.3 of the standard, it was proposed that labels for portable equipment should meet the requirements within the EU Noise Directive, which is based on a number of ISO standards. The portable equipment and machinery in the ANZ noise labelling and limit scheme were, by definition, not essentially stationary in nature. Hence, the labels for these articles do not necessarily need to meet the requirements within this standard.

4.3 Methods of Noise Control

4.3.1 Methods of Noise Control

The method of noise reduction can be classified into passive and active. The former ones are mostly used, which reduce the noise by converting sound energy into heat. Porous materials are good example of passive material, which reduces the sound energy by dissipating the heat energy due to the void structure. On the other hand, active materials work on the principle of active noise reduction. A second sound, with same amplitude but exactly opposite phase, is added to the first to cancel its effect. The two sound waves interact in a destructive interference, nullifying each other's effect. The active noise control is assisted by the use of digital signal processing. Programmed algorithms measure the wavelength of the noise and based on the results they generate an antiphase signal. The antiphase signal generates a sound proportionate to the original amplitude. The destructive interference reduces the effectiveness of the noise. Passive noise control depends on the noise reduction efficiency of the material, whereas active noise control depends on external power source. Active noise control is best suited for low frequency sounds. Passive noise control is more effective at higher sound frequencies.

The major areas where noise control is essential can be divided into four groups such as: residential, commercial, automotive and industrial [34, 35]. Noise in built environment such as home, work and leisure are generated from various sources such as televisions, household appliances, neighbours and industrial machineries. Residential noise control aims at controlling the exterior noises encountered by residents. The major area to be focused for this is the windows by the use of curtains. The curtains may be of honeycomb structures containing air or of heavy materials. The honeycomb structures are more effective than heavy materials. The limitation of use of curtains is the lack of proper edge sealing, although there is the scope of using hook and loop fasteners and adhesives. Significant noise reduction can be achieved using double-pane windows, where the external window is fixed and the internal window is displaceable. Furthermore, decorative textiles used in buildings such as upholstery, curtains, carpets and wall covering can assist in acoustic control.

Commercial places such as market, hospitals, auditoriums, schools and dining places need suitable noise control [35]. It is rather difficult to control the noise pollution in these places. While aiming for effective noise control in commercial places, the noise source, transmission path and the nature of building need to be considered. The acoustical design of buildings and isolating the noise sources from these places can help in achieving effective noise control. In the case of the impact of environmental noise such as road traffic on commercial places, reducing the number of vehicles and the speed can help. Increasing the distance between the source and the receiver can help in reducing the outdoor noise level. The increased distance helps in the spreading of the noise in a larger area, absorption by the ground and the atmosphere.

Automobiles produce a variety of noises depending on the speed and surrounding conditions. Automotive noise control aims at preventing the external noise of tyre, engine and exhaust reaching at the interior [35, 36]. The noise generated in vehicles can be controlled by altering the surface texture of the road, limiting speed of vehicles, installing barriers, reducing the number of heavy vehicles and optimum tyre design. Failing to tackle noise from vehicles at source requires national governments, local authorities and homeowners to install noise barriers or sound insulation to homes and public buildings. Quieter vehicles reduce the need for expensive noise abatement and would increase property values, since homes in noisy areas are less attractive to potential buyers [37]. The limitations in automotive noise control lie in the weight and thickness of the material used in order to make the automobile lighter. However, significant results can be achieved to reduce the vehicular noise by the combinations of sound absorbers, barriers and dampers. The other areas to be focused include use of noise barriers, improved road design, speed limitations, truck restrictions, urban planning and surface pavement selection. Controlling the speed of vehicles is one of the major factors for effective noise reduction. Vehicles moving at a speed lower than 60 km/h generate very low noise. Above this limit, the noise level almost doubles with increase in the speed by 8 km/h.

Industrial noise can be controlled by redesigning of machineries, isolating the machineries in separate chambers, using barriers and shock mounted assemblies [34]. Approaches such as buying quieter machines, acoustic quieting, strict regulations and increasing the distance between the source and receiver are some of the approaches that can help in controlling the industrial noise pollution.

Special designed earplugs can be used by the workers to protect hearing loss when they work in high noise levels in excess of 85 dB for 8 h or 103 dB for 7 min. Earplugs are rated as per their ability to reduce the noise level and given a noise reduction rating (NRR) following ANSI S3.19-1974. The prolonged use can lead to allergic reactions, cause pain in the ear if pushed against the eardrum. Hence, the use of custom-shaped earplugs can help to avoid these problems. Similarly acoustic earmuffs can be used for hearing protection of the workers.

The noise from aircrafts and spacecrafts can be reduced by designing quieter jet engines, altering flight paths and the flight timings [36, 38]. The last two approaches are beneficial for the people living close to the airport. The designers, manufacturers and operators of aircraft are trying to develop quieter aircrafts and practice better operating procedures to reduce the noise level. The newly designed high-bypass turbo engines are found to be quieter than the low-bypass turbofan and turbojet engines. Modelling and simulation can be used to study the effects of aircraft type, and flight patterns on building structure and sound level. Building modifications such as roof upgradation, fireplace baffling, window glazing improvement, caulking construction can be done to reduce the problems. Night flying restrictions can be applied to reduce the noise level at night [39].

4.3.2 Room Acoustic

Room acoustic deals with the creation, propagation, perception, measurement and modelling inside enclosed places such as auditoriums, theatres, offices and seminar halls [34, 40]. The most important parameter that describes room acoustic is the reverberation time (RT). The level of understanding as well as the distribution of sound depends on RT. For proper understanding, the RT should be balanced over the frequency spectrum. Too long RT value makes the understanding difficult as one sound or word cannot be uniquely distinguished from the other. Similarly, too short RT value makes sound levels too low at a distance. Very high RT may lead the sound level to rise that can lead to the risk of impaired hearing. Architectural acoustics deals with the engineering of achieving a good sound control inside a building such as concert hall, theatre and restaurant. It can also involve improving the quality of sound in a recording studio or concert hall.

4.3.3 Automotive Noise Control

Noise control in vehicles is essential for manufacturers to gain market competitiveness and price appreciation [37, 41]. The main method employed to achieve this is using sound absorbing materials attached to various components such as headliners, door panels, floor coverings, dash boards and trunk liners. Automotive manufacturers also try to reduce the exterior noise from exhaust, engine and tyres.

This is also a challenging area as there are limitations on the thickness of the material used. The nature of the vehicle, speed as well as the driving conditions can alter the amount of noise. The effectiveness of noise control can be improved by the combined use of sound absorbers, barriers, baffles and dampers. A proper understanding is necessary to achieve effective automotive noise control using lighter material with reduced thickness.

4.4 Materials Used for Noise Control

Although, almost all materials possess some amount of sound absorption property, acoustic materials are those that can absorb the majority of the sound energy impinged on them [42, 43]. Acoustic materials can be used to control and reduce the noise levels from various sources. These materials absorb and dissipate the energy converting some into heat when sound travels through them. Acoustic materials can be of two types such as (a) noise absorption type and (b) noise reduction type. The former class of materials works by suppressing the sound, whereas the latter class works by reducing the sound energy when it passes through them. Noise reduction coefficient (NRC) and noise attenuation are the parameters that describe the ability of these materials to reduce the noise. NRC represents the amount of sound energy absorbed by a material when sound wave strikes a particular surface, which ranges from 0 to 1. A NRC value of 0 indicates perfect reflection; whereas a NRC of 1 indicates perfect absorption. NRC is measured by acoustic instruments using frequencies of 250, 500, 1000 and 2000 Hz.

Acoustic materials reduce the energy of sound waves before it is reached with the receptor. Several methods can be used to reduce the sound such as: (a) use of acoustic barriers that can absorb sound energy, (b) increasing the distance between the source and receptor, (c) use of sound baffles and (d) use of antinoise sound generators. All these methods are based on the following four principles:

- Sound absorption: Several materials can absorb sound energy and covert into heat. The porous materials are best suited for this purpose. The common materials used for sound absorption include fibrous structures, open cell foams, fibre glass and lead-based tiles.
- Sound damping: In damping process, the energy of sound is dissipated as the oscillation gradually reduces. In damping process, the vibration energy is extracted from a thin sheet that is dissipated in the form of heat. Damping is applicable to large vibrating surfaces. An example of damping material is deadened steel.
- Sound insulation: This principle applies barriers to prevent the transmission. Materials with high density such as concrete, metal, thick glass and brick are used for this purpose.

- Vibration isolation: the transmission of sound energy is prevented from a sound source by introducing a physical break. The materials used for this purpose are cork, springs and rubber mounts.

Many acoustic materials work on the principle of sound absorption to reduce the sound energy. These materials, either fibrous or porous are generally resistive in nature or may be reactive resonators in some cases [44]. The resistive materials include foams, felt, fibrous glass, nonwovens and mineral wool; whereas the resonators include sintered metal and masonry blocks. These products provide varying degree of insulation depending on the sound frequency by reducing the amplitude of the reflected waves. The sound absorption characteristics of a material depend on its thickness, elastic modulus, flow resistance and porosity. Flexible textile structures are better suited for low frequency sounds than the stiff products of equivalent thickness [45, 46]. For low frequency sounds, the higher is the thickness; the better is the sound absorption up to a certain limit.

Porous materials or resonant structures can be used as sound absorbers [47]. For sound absorption, the main mechanisms are acoustic impedance matching the sound absorbers' boundary and acoustic energy dissipation inside the absorbers. In porous absorbers, sound propagation occurs in a network of interconnected pores, and acoustic energy is dissipated by viscous and thermal effects. Resonant absorbers have two common forms: membrane/panel absorbers and Helmholtz absorbers. Porous absorbers are effective for broadband from mid- to high frequency rages, whereas the resonant absorbers are usually only effective in a narrow tuneable low frequency band.

4.4.1 Porous Materials

Various materials can transform the energy of sound wave to heat when the sound wave is passed through these materials [47, 48]. However, the porous materials with porosity of >90 % allow sound to enter into the structure. A series of interlocking pores in the porous material help in converting the sound energy into heat. The best example of a porous structure includes open cell foam. The closed pores are less effective in sound absorption due to less frictional resistance. A porous material can more effectively absorb low frequency sound if an air space is present behind the material.

When sound energy enters into a porous material, air molecules start oscillating in the interstices of the porous material at same frequency as the sound wave. Frictional losses and momentum losses are involved with the oscillation. The former is associated with the oscillation whereas the latter is associated with the flow direction change of sound waves, in addition to the expansion and contraction phenomenon. Air molecules inside the pores undergo periodic compression and relaxation, which lead to the generation of heat.

At low sound frequencies, the heat is exchanged in isothermal conditions, whereas at high frequencies compression occurs adiabatically. The loss of sound energy due to heat exchange occurs in the frequency range between the isothermal and adiabatic compression. If the propagation of sound is parallel to the fibre plane, this loss is high. Hence, the loss of energy by the sound wave when the sound passes through the acoustic material can be attributed to: (a) frictional losses, (b) momentum losses and (c) temperature fluctuations.

The sound in low frequency is rather difficult to absorb due to its long wave length and we are less susceptible to these sounds [47]. The denser is the material, the lower is the sound absorption. Hence, the porous fibrous structures serve better in sound absorption compared to solid panels.

The porous materials used for acoustic purposes can be classified as porous foam or fibrous structure [48]. The fibrous structures are prepared by methods such as weaving, nonwoven and composite using fibres such as polyester, polypropylene, rockwool and glass, which are discussed in the next section. Sound absorptive materials are different from sound barriers. Sound absorbers work by absorption of the sound wave, whereas the barriers work on transmission loss. Generally, sound absorbers are porous and light but barriers are solid and heavy. A good sound absorber may provide poor barrier. Unlike barriers, the mass of an absorptive material bears no direct relation on acoustic absorption [49]. Sound absorbers always work in combination with the barriers as some noise can pass through the porous materials. In this case, the barrier can convert the sound energy into heat. Hence, the barriers reduce the strength of reflected noise and prevent sound build-up in enclosed places.

Although there are several light-weight porous acoustic materials available, the effectiveness of these materials at low to medium frequency sound is relatively poor. The application of nanotechnology in many other fields has shown promising results for low frequency sound absorption [50–53]. The application of innovative nanofibres and the porous structures derived from them may effectively work for the noise control at lower frequencies.

4.4.2 Virgin Materials

Various flexible and porous textile structures produced using textile fibres can be used for acoustic applications. A textile structure is manufactured from natural, synthetic, regenerated or the blends of these fibres using techniques of weaving, knitting or nonwoven. Depending on the manufacturing method, the acoustic properties of the textile material can vary. The nature of the pores in the structure and the fibre configuration affects the acoustic properties. Among the textile structures, nonwovens are preferred for acoustic application. However, they lack the aesthetic appeal. Hence, they are draped with the woven fabrics to produce pleasing appearance.

Textile structures are thinner, lighter and higher sound absorbent for specific applications. By altering the structure, it is possible to achieve similar sound absorption using a lighter textile structure instead of a bulkier material. However, there are certain limitations such as low density of a textile structure, which is a disadvantage when sound diffusion is considered. It lacks the stiffness and air tightness. In addition, textiles are not homogeneous, as the properties vary in different angles. The design of 3D structures can improve these parameters for effective diffusion of sound waves of high frequencies. This also applies to adsorption depending on the frequency range.

Majority of the fibres used in acoustic applications are natural fibres although there has been some recent work on the use of natural fibre blends with synthetics [43, 44]. Kenaf and ramie are the main natural fibres used for automotive interiors due to their high tenacity. Other natural fibres such as coir produced from coconut husk, cotton, jute, flax and hemp are also potential material for acoustic absorption [54]. Parameters such as fibre size, geometry, physical properties, mechanical properties and durability are considered while selecting the fibres. The textile structures are used in various acoustic applications such as upholstery in auditoriums and concert halls; automotive insulations and acoustic panels for work stations. They are also used as interior lining for ducts, airplanes, automotive and apartments; insulating material for appliances and enclosures for noisy equipment.

4.4.3 Recycled Materials

Environmental concerns have led to the use of biodegradable and recyclable materials in many products [55]. Natural fibres are biodegradable, renewable, easily available and eco-friendly, for which they are now increasingly used in noise control. They can be recycled and reused at the end-of-life. The recycled products of the natural fibres can effectively absorb the sound by energy dissipation, when placed between the source and receiver [49, 56, 57].

The nonwoven industries generate a large quantity of waste that causes environmental challenges. Furthermore, the production and consumption of apparel, home textiles and industrial textiles has increased by manifolds [58]. The amount of waste generated from these products is alarming to the ecosystem. Hence, the concept of recycle and reuse plays an important role in reducing the environmental loading and helps in the effective use of resources. Whether, virgin or recycled materials used in the construction, the sound absorption efficiency depends on the porous structure and thickness of the textile material.

Several researches have been done on recycled materials for acoustic applications [57, 59–63]. For example, Jordeva et al. [64] designed an acoustic panel from polyester apparel cutting waste and evaluated the sound insulation properties of the panel. It was observed that the acoustic material produced from apparel cutting waste has good sound absorption compared to standard sound insulators. The sound absorption achieved was in the range of 54.7–74.7 %, when tested in the frequency range of 250–2000 Hz. This research established a method to lower the environmental loading by recycling apparel cuttings waste for acoustic absorption in internal walls. Furthermore, wastes of natural materials such as industrial tea-leaf fibre [65]), agricultural waste [66], rubber particles [61] and recycled [65] foam also have been the area of research in the form of composites in acoustic applications.

4.4.4 Composites

The composites of natural fibres (with thermoplastics or thermosetting) are used by several car manufacturers and in other areas for acoustic purposes [43]. The composites possess excellent properties such as low weight, high strength and stiffness; which enable them for several areas of technical applications including acoustics. Due to the increasing environmental concerns and demands of legislative authorities, the use of traditional composites using synthetic fibre materials such as glass, carbon, or aramid is decreasing. On the other hand, due to eco-friendly and biodegradability, the composites of natural fibres are being widely used in many technical applications. Furthermore, there are other benefits such as reduced cost and improved safety. Natural fibre reinforced composites offer increased recycling capabilities over conventional polymer composites. Several researches have been done on the composite materials for acoustic applications [67–70].

5 Book Contents

Chapter "Acoustic Textiles: An Introduction" gives an introduction to acoustic textiles. This chapter provides some statistical data on the global usage of acoustic textiles and the reasons for the increased use. This chapter also covers the sources of noise and the effects of noise pollution on human beings and animals, covers the brief history and need for acoustic textiles. The regulations followed in many countries of the world are also highlighted. Various methods of noise control and materials used for the same are also being discussed. Furthermore, the standards used to evaluate acoustic textiles and governing bodies around the world for acoustics are also covered.

In Chapter "Basics of Acoustic Science", the basic wave phenomena related to sound propagation such as acoustic reflection, absorption and scattering are introduced first, and then the corresponding parameters such as transmission loss, absorption coefficient, scattering coefficient and flow resistance are explained.

Chapter "Principles of Sound Absorbers" deals with the principles of sound absorbers and the factors that affect acoustic absorption. This chapter also deals with the different types of sound absorbers such as porous materials and resonant structures and the differences between them. The applications of sound absorbers are focused on the discussion of reverberation time control and sound pressure level control in rooms. The contents of this chapter are focused on the fundamental aspects of sound absorption and absorbers to serve as an introduction to students and designers with basic knowledge in mathematics and physics. This chapter also describes the factors affecting sound absorption, which are related to the material properties, design of the acoustic material and nature of arrangement discussed in detail. Finally, specific discussions are given to acoustic textiles and its recent developments.

Chapter "Materials used for Acoustic Textiles" describes the materials used for acoustic textiles. Different types of natural, synthetic and recycled raw materials used in the production of acoustic textiles are discussed. Different types of porous absorbing materials as well as panel absorbing materials are discussed. Acoustic properties of materials coated, sprayed or embedded with chemicals and fillers are also covered. Furthermore, various chemicals used for coating or finishing of acoustic textiles are also being discussed. This chapter also describes the current application of nanomaterial to improve the acoustic absorption particularly for absorbing low and medium frequencies. Some of the recent trend in developing hybrid type of acoustic absorbers, where acoustic property is effectively controlled with the help of active as well as passive absorbers is also covered. In addition, new types of panel absorbers are also discussed.

Chapter "Manufacturing Methods for Acoustic Textiles" focuses on the manufacturing methods used for acoustic textiles. Various production methods used to produce woven, nonwoven and composites for acoustic applications is discussed. This chapter also highlights the application of 3D materials in acoustic textiles. Furthermore, this chapter focuses on fibrous mats, needle punched nonwovens and polymeric composites made of nonwovens that are widely researched textile materials nowadays. The approach of combining two or more textile structures also used to counter the noises at low frequencies, is also covered in this chapter. Final part of the chapter deals with emerging techniques that are reported by very few researchers and may lead to more interesting outcomes in coming years.

Chapter "Acoustical Test Methods for Nonwoven Fabrics" highlights the evaluation methods for important physical properties of fabrics for acoustical applications such as areal density (mass), volumetric density, porosity, particle size distribution, tortuosity and thickness. Furthermore, the mechanical parameters those are related to acoustic properties of nonwoven fabric such as tensile properties and Young's Modulus are also discussed. Acoustical absorption properties determined using sound absorption coefficient, sound transmission loss, noise reduction coefficient, sound impedance and airflow resistivity are also discussed. In addition, important standard or nonstandard test methods related to acoustic properties of nonwoven fabrics are described. A special emphasis has been given to impedance tubes and reverberation rooms used for most of the acoustical test methods.

Chapter "Applications of Acoustic Textiles in Automotive/Transportation" aims to discuss those textile developments that are used mainly to provide noise isolation and sound absorption in different means of transport such as aircrafts, vehicles, ship, spacecraft and other transport areas. Acoustic textiles used to control noise in vehicles must provide airborne transmission reduction, damping and sound absorption. The use of acoustic textiles for noise control of main sources in cars is discussed. The final section is devoted to further applications of acoustic textiles in aircraft, trains, ships and spacecrafts.

Chapter "Application of Acoustic Materials in Civil Engineering" describes the application of acoustic materials in civil engineering. Various applications of acoustic textiles in auditoriums, halls, and seminar rooms, libraries, courts and wherever sound insulation needed is covered in this chapter. Different materials such as acoustic wood panels and wood composites; autoclaved aerated concrete, acoustic foam products; woven, nonwoven and fibre retrofitting for walls and noise absorbing road surfaces have also been discussed.

Chapter "Design of Acoustic Textiles: Environmental Challenges and Opportunities for Future Direction" highlights the future directions and design challenges faced for acoustic materials. This chapter investigates acoustic textiles in terms of their environmental impact. This chapter also deals with the design stage referring to life cycle assessment methods. The chapter examines the effects of manufacturing stage of the acoustic textiles. Furthermore, this chapter also reviews different materials used for sound absorption and ecolabels. Finally, this book also considers recycling issues and end-of-life concerns of acoustic textiles.

6 Standards and Governing Bodies Relating to Acoustic Materials

6.1 Standards

There is a significant influence of regulations and standards on design of quiet equipment so as a quiet work place. Furthermore, these standards can be used for the evaluation of acoustic structures, materials and measure noise emission levels. Regulations such as European Directive 86/188 (Art. 8, Par. 1b) and 89/392 (Annex I, Sect. 1.7.4f) exist, which specify methods to evaluate the noise emitted by machineries. The specifications can help the buyers to select the quieter machineries. Some of the standards followed around the globe are discussed in Table 3.

Standards such as ISO 3740, ISO 3744 and ISO 3746 are used to determine the sound power using the enveloping measurement surface procedure. ISO 11201, ISO 11202, ISO 11203 and ISO 11204 are used to determine the emission sound pressure level, which are complement to ISO 3744. As the emission sound pressure level is considered as a criterion for the type and scope of the noise declaration, these standards can be used for the precise determination. ISO 11203 specifies procedures for calculating the emission sound pressure level from the sound power level, whereas ISO 11204 ensures the most correct possible environmental correction for a measuring point. ISO 11203 is very effective for smaller machines even for household appliances. Although ISO 11201 and ISO 11202 contain simplified procedures, they are only applicable to a limited extent. ISO 4871 specifies the type of noise declaration and a simplified verification procedure for machinery and equipment.

Several standards exist relating to the determination, declaration and verification of noise emission, which enables to evaluate the noise level and quality. ISO/TR 11,690-3 recommends the means of designing a low-noise workplace in machinery environments. ISO/TR 11688 (Part 1 and part 2) are devised to reduce noise of

Table 3 Various standards relating to acoustic regulations, test methods and materials

ASTM C 522-03	Standard test method for airflow resistance of acoustical materials
ASTM C 634-08	Standard terminology relating to building and environmental acoustics
ASTM E2611-09: 2009	Standard test method for measurement of normal incidence sound transmission of acoustical material based on the transfer matrix method
ASTM E1050-10: 2010	Standard test method for impedance and absorption of acoustical materials using a tube, two microphones and digital frequency analysis system
BS EN 292-1:1991	Safety of machinery. Basic concepts, general principles for design. Basic terminology, methodology
EN 292-2:1991	Safety of machinery. Basic concepts, general principles for design. Technical principles and specification
BS EN 414:2000	Safety of machinery. Rules for the drafting and presentation of safety standards
ISO 11654: 2002	Acoustics—sound absorbers for use in buildings—rating of sound absorption
ISO 11546-1:2009	Acoustics—determination of sound insulation performances of enclosures—part 1: measurements under laboratory conditions
ISO 11546-2:2009	Acoustics—determination of sound insulation performances of enclosures—part 2: measurements in situ
ISO 354:2003	Acoustics—measurement of sound absorption in a reverberation room
ISO 11957:2009	Acoustics—determination of sound insulation performance of cabins (laboratory and in situ measurements)
ISO 7235:2003	Acoustics—laboratory measurement procedures for ducted silencers and air-terminal units—insertion loss, flow noise and total pressure loss
ISO 7574-1:1985	Acoustics—statistical methods for determining and verifying stated noise emission values of machinery and equipment—part 1: general considerations and definitions
ISO 7574-2:1985	Acoustics—statistical methods for determining and verifying stated noise emission values of machinery and equipment—part 2: methods for stated values for individual machines
ISO 7574-3:1985	Acoustics—statistical methods for determining and verifying stated noise emission values of machinery and equipment—part 3: simple (transition) method for stated values for batches of machines
ISO 7574-4:1985	Acoustics—statistical methods for determining and verifying stated noise emission values of machinery and equipment—part 4: methods for stated values for batches of machines
ISO 11691:2009	Acoustics—measurement of insertion loss of ducted silencers without flow—laboratory survey method
ISO 11820:1996	Acoustics—measurements on silencers in situ
ISO 11821:1997	Acoustics—measurement of the in situ sound attenuation of a removable screen
ISO 13472-2: 2010	Acoustics—measurement of sound absorption properties of road surfaces in situ—part 2: spot method for reflective surfaces
ISO 10844: 2010	Acoustics—specification of test tracks for measuring noise emitted by road vehicles and their tyres
ISO 10534-1:1996	Acoustics—determination of sound absorption coefficient and impedance in impedance tubes—part 1: method using standing wave ratio
ISO 10534-2: 1998	Acoustics—determination of sound absorption coefficient and impedance in an impedance tube—part 2: transfer-function method
ISO 14163:1998	Acoustics—guidelines for noise control by silencers
ISO 10053:1991	Acoustics—measurement of office screen sound attenuation under specific laboratory conditions

(continued)

Table 3 (continued)

ASTM C 522-03	Standard test method for airflow resistance of acoustical materials
ISO 4869-1:1990	Acoustics—hearing protectors—part 1: subjective method for the measurement of sound attenuation
ISO 4869-2:1994	Acoustics—hearing protectors—part 2: estimation of effective a-weighted sound pressure levels when hearing protectors are worn
ISO 4869-2:1994	Acoustics—hearing protectors—part 3: measurement of insertion loss of ear-muff type protectors using an acoustic test fixture
ISO 4871:1996	Acoustics—declaration and verification of noise emission values of machinery and equipment
ISO 3740:2000	Acoustics–determination of sound power levels of noise sources—guidelines for the use of basic standards
ISO 3744:2010	Acoustics—determination of sound power levels and sound energy levels of noise sources using sound pressure—engineering methods for an essentially free field over a reflecting plane
ISO 3746:2010	Acoustics—determination of sound power levels and sound energy levels of noise sources using sound pressure—survey method using an enveloping measurement surface over a reflecting plane
ISO 11201:2010	Acoustics—noise emitted by machinery and equipment—determination of emission sound pressure levels at a work station and at other specified positions in an essentially free field over a reflecting plane with negligible environmental corrections
ISO 11202:2010	Acoustics—noise emitted by machinery and equipment—determination of emission sound pressure levels at a work station and at other specified positions applying approximate environmental corrections
ISO 11203:2009	Acoustics—noise emitted by machinery and equipment. Determination of emission sound pressure levels at a work station and at other specified positions from the sound power level
ISO 11204:2010	Acoustics—noise emitted by machinery and equipment—determination of emission sound pressure levels at a work station and at other specified positions applying accurate environmental corrections
ISO/TR 11688-1:1995	Acoustics—recommended practice for the design of low-noise machinery and equipment—part 1: planning
ISO/TR 11688-2:1995	Acoustics—recommended practice for the design of low-noise machinery and equipment—part 2: introduction to the physics of low-noise design
ISO 11689:1996	Acoustics—procedure for the comparison of noise emission data for machinery and equipment
ISO 11690-1:1996	Acoustics—recommended practice for the design of low-noise workplaces containing machinery—part 1: noise control strategies
EN ISO 11690-2:1996	Acoustics—recommended practice for the design of low-noise workplaces—part 2: noise control measures
ISO/TR 11690-3:1997	Acoustics—recommended practice for the design of low-noise workplaces containing machinery—part 3: sound propagation and noise prediction in workrooms
ISO 11691:1995	Acoustics—measurement of insertion loss of ducted silencers without flow—laboratory survey method

machines, which deal with the planning and design of low-noise machines. ISO 11689 is used for the collection and evaluation of emission data for machineries. Other standards for noise control materials and devices include:

- ISO 11546, ISO 11957 (for noise control of enclosures),
- ISO 11654, ISO 10534 (for noise absorbers),
- ISO 7235, ISO 11691, ISO 11820, ISO 14163 (for silencers),
- ISO 11821, ISO 10053 (for noise baffles).

6.2 Professional Societies

There are several organizations/societies dealing with the regulations of sound, policy formulation, standard preparation and evaluation of acoustic materials. These organizations also focus on the goal of increasing the knowledge of acoustics and to promote its practical applications. They may deal with the diverse fields of acoustics such as acoustic physics, oceanography, biology, engineering, architecture, noise, psychology, music, speech and hearing. A list of various organizations around the world is mentioned below.

- Acoustical Society of America (ASA)
- Audio Engineering Society (AES, (USA))
- American Society of Mechanical Engineers, Noise Control and Acoustics Division (ASME-NCAD)
- American Institute of Aeronautics and Astronautics, Aeroacoustics (AIAA)
- Australian Acoustical Society (AAS)
- Acoustical Society of New Zealand
- Acoustical Society of Thailand
- Acoustical Society of Argentina
- Acoustical Society of Chile
- Acoustical Society of Croatia
- Acoustical Society of China
- Acoustical Society of Denmark
- Acoustical Society of Egypt
- Acoustical Society of Finland
- Acoustical Society of Hungary
- Acoustical Society of India
- Acoustical Society of Japan
- Acoustical Society of Korea
- Acoustical Society of the Netherlands
- Acoustical Society of New Zealand
- Acoustical Society of Nigeria
- Acoustical Society of Norway
- Acoustical Society of Sweden

- American Academy of Audiology
- American Speech Language Hearing Association
- Audio Engineering Society
- Austrian Acoustics Association
- Belgian Acoustical Association
- Canadian Acoustical Association
- Catgut Acoustical Society
- Czech Acoustic Society
- European Acoustics Association
- French Acoustical Society (Société Française d'Acoustique—SFA)
- German Acoustical Society (Deutsche Gesellschaft für Akustik)
- Hellinic Institute of Acoustics
- Hong Kong Institute of Acoustics
- Iberoamerican Federation of Acoustics (FIA)
- Icelandic Acoustical Society
- Institute of Acoustics
- Institute of Mexican Acoustics
- Institute of Electrical and Electronics Engineers (IEEE)
- Institute of Acoustics (IoA, UK)
- Institute of Noise Control Engineering
- International Commission for Acoustics
- International Bio-Acoustic Council
- International Institute of Noise Control Engineering
- International Commission for Acoustics (ICA)
- International Institute of Acoustics and Vibration
- International Speech Communication Association
- Israeli Acoustical Association
- Italian Acoustical Society
- Italian Association of Laser Velocimetry and Non-Invasive Diagnostics
- Latvian Acoustical Society
- Middle East Acoustic Society
- National Council of Acoustical Consultants
- National Hearing Conservation Association
- Polish Acoustical Society
- Noise Pollution Clearinghouse
- Portuguese Acoustical Society
- Professional Lighting and Sound Association (PLASA)
- Romanian Acoustical Society
- Russian Acoustical Society
- Slovak Acoustical Society
- Slovenian Acoustical Society
- Sociedade Brasileira de Acustica (SOBRAC)
- Society of Acoustics (Singapore)
- South African Acoustics Institute
- Spanish Acoustical Society (Sociedad Española de Acústica)

- Speech-Language and Audiology Canada
- Swiss Acoustical Society
- Turkish Acoustical Society
- Ultrasonic Industry Association
- World Forum for Acoustic Ecology

6.3 List of Journals and Global Conferences

6.3.1 List of Journals

1. IEEE Transactions on Audio, Speech and Language Processing
2. Ultrasonics Sonochemistry
3. Journal of Sound and Vibration
4. Journal of Physics D—Applied Physics
5. Journal of Vibration and Acoustics, Transactions of the ASME
6. IEEE Transactions on Ultrasonics, Ferroelectrics, and Frequency Control
7. Synthesis Lectures on Speech and Audio Processing
8. Ultrasonics
9. Journal of the Acoustical Society of America
10. Journal of Computational Acoustics
11. Applied Acoustics
12. Acta Acustica united with Acustica
13. Zhendong yu Chongji/Journal of Vibration and Shock
14. International Journal of Aeroacoustics
15. Medical Ultrasonography
16. Zhendong Gongcheng Xuebao/Journal of Vibration Engineering
17. Phonetica
18. Zhendong Ceshi Yu Zhenduan/Journal of Vibration, Measurement and Diagnosis
19. Noise Control Engineering Journal
20. Archives of Acoustics
21. Acoustical Physics
22. Acoustics Australia
23. Acoustical Science and Technology
24. International Journal of Acoustics and Vibrations
25. Building Acoustics
26. Yadian Yu Shengguang/Piezoelectrics and Acoustooptics
27. Journal of Low Frequency Noise Vibration and Active Control
28. Advances in Acoustics and Vibration
29. Larmbekampfung
30. Romanian Journal of Acoustics and Vibration
31. Noise and Vibration Worldwide

32. Eurasip Journal on Audio, Speech, and Music Processing
33. Shengxue Xuebao/Acta Acustica
34. Synthesis Lectures on Antennas
35. Canadian Acoustics—Acoustique Canadienne
36. Akustika
37. Acoustics Bulletin
38. IUTAM Bookseries
39. Fluid Dynamics
40. Journal of Voice
41. Physics of Fluids
42. Sound and Vibration magazine
43. Technical Acoustics
44. The Journal of the Acoustical Society of America
45. Theoretical and Computational Fluid Dynamics.

6.3.2 List of Global Conferences

1. AIAA/CEAS Aeroacoustics Conference
2. Summer School on Acoustics
3. EAA (European Acoustics Association) Conferences
4. Euronoise
5. Baltic Nordic Acoustic Meeting
6. Lasers and Photonics Congress
7. International Conference on the Effects of Noise on Aquatic Life
8. Optical and Ultrasound imaging—OPUS Innovative Multiphysics Coupling for Biomedical Imaging
9. International Congress on Sound and Vibration
10. International Conference "Material—Acoustics—Place"
11. International Commission on Biological Effects of Noise (ICBEN)
12. Ergonomics
13. Acoustic and Environmental Variability, Fluctuations and Coherence
14. Acoustics
15. Mediterranean International Workshop on Photoacoustic and Photothermal Phenomena
16. "Open GIS and Open Data for environmental noise assessment" Session at OGRS
17. FFT Acoustic Simulation Conference and Actran Users' Meeting 2016
18. AAA (Austrian Acoustics Association) Congress on Sound and Vibration
19. Noise and Vibration Engineering Conference
20. International Ultrasonics Symposium (IUS)
21. International Symposium on Music and Room Acoustics
22. International Conference on Digital Audio Effects
23. VIBROENGINEERING Conference

24. IEEE International Ultrasonics Symposium
25. CBEN Congress on Noise as a Public Health Problem
26. Asia Pacific Symposium on Cochlear Implants and Related Sciences
27. Underwater Acoustics Conference.

7 Conclusion

Noise pollution is an ever increasing problem around the world. The increased number of vehicles, construction work, large buildings, home appliances, rail transport, airplane and industrial machineries has led to increased noise pollution. Sound absorbing materials are used to control the noise pollution. This chapter has discussed various sources of noise pollution, the impact of noise on the health of humans and animals, global usage of acoustic materials, legality for noise control in many countries, method and materials used for noise control, standards used for acoustic materials and various organizations dealing with acoustics. It is essential to reduce the amount of noise by acoustically designing buildings, structures and roads. Furthermore, the implementation of stricter regulations for noise control by vehicles, aircrafts, industries and other social activities; designing of quieter machinery and equipment; and development of new standards for measurement of noise level and quality can help in achieving reduced noise level. A more effective implementation of the noise standards would provide a significant improvement of the health and quality of life of citizens.

References

1. Morfey CL (2000) Dictionary of acoustics. Academic Press
2. Raichel DR (2006) The science and applications of acoustics. Springer Science & Business Media
3. Franssen E, Van Wiechen C, Nagelkerke N, Lebret E (2004) Aircraft noise around a large international airport and its impact on general health and medication use. Occup Environ Med 61(5):405–413
4. Pandya G (2003) Assessment of traffic noise and its impact on the community. Int J Environ Stud 60(6):595–602
5. Garg N, Gupta V, Vyas R (2007) Noise pollution and its impact on urban life. J Environ Res Dev 2:290–296
6. Goines L, Hagler L (2007) Noise pollution: a modern plague. South Med J Birmingham Alabama 100(3):287
7. Schmidt FP, Basner M, Kröger G, Weck S, Schnorbus B, Muttray A, Sariyar M, Binder H, Gori T, Warnholtz A (2013) Effect of nighttime aircraft noise exposure on endothelial function and stress hormone release in healthy adults. Eur Heart J eht269
8. Stansfeld SA, Berglund B, Clark C, Lopez-Barrio I, Fischer P, Öhrström E, Haines MM, Head J, Hygge S, Van Kamp I (2005) Aircraft and road traffic noise and children's cognition and health: a cross-national study. The Lancet 365(9475):1942–1949

9. Dalton BH, Behm DG (2007) Effects of noise and music on human and task performance: a systematic review. Occup Ergon 7(3):143–152
10. Mostafapour SP, Lahargoue K, Gates GA (1998) Noise-induced hearing loss in young adults: the role of personal listening devices and other sources of leisure noise. Laryngoscope 108(12):1832–1839
11. Wiwanitkit V (2010) Noise and sports events. SAMJ S Afr Med J 100(6):334
12. Rylander R, Lundquist B (1996) Annoyance caused by noise from heavy weapon shooting ranges. J Sound Vib 192(1):199–206
13. Pathak V, Tripathi B, kumar Mishra V (2008) Evaluation of traffic noise pollution and attitudes of exposed individuals in working place. Atmos Environ 42(16):3892–3898
14. Sorvari J, Antikainen R, Pyy O (2006) Environmental contamination at finnish shooting ranges—the scope of the problem and management options. Sci Total Environ 366(1):21–31
15. Crocker MJ, Li Z, Arenas JP (2005) Measurements of tyre/road noise and of acoustical properties of porous road surfaces. Int J Acoust Vib 10(2):52–60
16. Parris KM, Schneider A (2009) Impacts of traffic noise and traffic volume on birds of roadside habitats. Ecol Soc 14(1):29
17. Mbuligwe SE (2004) Levels and influencing factors of noise pollution from small-scale industries (SSIs) in a developing country. Environ Manage 33(6):830–839
18. Babisch W (2008) Road traffic noise and cardiovascular risk. Noise Health 10(38):27
19. Babisch W, Beule B, Schust M, Kersten N, Ising H (2005) Traffic noise and risk of myocardial infarction. Epidemiology 16(1):33–40
20. Sørensen M, Hvidberg M, Andersen ZJ, Nordsborg RB, Lillelund KG, Jakobsen J, Tjønneland A, Overvad K, Raaschou-Nielsen O (2011) Road traffic noise and stroke: a prospective cohort study. Eur Heart J 32(6):737–744
21. Stansfeld S, Haines M, Brown B (2000) Noise and health in the urban environment. Rev Environ Health 15(1–2):43–82
22. Nissenbaum MA, Aramini JJ, Hanning CD (2012) Effects of industrial wind turbine noise on sleep and health. Noise Health 14(60):237
23. Moudon AV (2009) Real noise from the urban environment: how ambient community noise affects health and what can be done about it. Am J Prev Med 37(2):167–171
24. Passchier-Vermeer W, Passchier WF (2000) Noise exposure and public health. Environ Health Perspect 108(Suppl 1):123
25. Morrell S, Taylor R, Lyle D (1997) A review of health effects of aircraft noise. Aust N Z J Public Health 21(2):221–236
26. Hurtley C (2009) Night noise guidelines for Europe. WHO Regional Office Europe
27. Pepper CB, Nascarella MA, Kendall RJ (2003) A review of the effects of aircraft noise on wildlife and humans, current control mechanisms, and the need for further study. Environ Manage 32(4):418–432
28. Evans E, England G (2000) Joint interim report Bahamas Marine mammal stranding. US Department of Commerce
29. Zannin PHT, Ferreira AMC, Szeremetta B (2006) Evaluation of noise pollution in urban parks. Environ Monit Assess 118(1–3):423–433
30. Noise Control Act (1972) Environmental protection agency. Public Law 92–574. USA
31. Noise Control Act (1976) Environmental protection agency. Amendment, Public Law 94–301. USA
32. Noise Control Act (1978) Environmental protection agency. Amendment, Public Law 95–609. USA
33. Noise Control Act (1988) Environmental protection agency. Amendment, Public Law 100–418
34. Crocker MJ (2007) Handbook of noise and vibration control. Wiley
35. Shoureshi R, Knurek T (1996) Automotive applications of a hybrid active noise and vibration control. Control Syst IEEE 16(6):72–78
36. Rao MD (2003) Recent applications of viscoelastic damping for noise control in automobiles and commercial airplanes. J Sound Vib 262(3):457–474

37. Karbon KJ, Singh R (2002) Simulation and design of automobile sunroof buffeting noise control. In: Eighth AIAA/CEAS aeroacoustics conference
38. Dimino I, Aliabadi F (2015) Aircraft Noise Control. In: Active Control of Aircraft Cabin Noise. World Scientific, pp 1–15
39. Night flying restrictions at Heathrow, Gatwick and Stansted Airports (2006)
40. Fry A (2013) Noise Control in Building Services: Sound Research Laboratories Ltd. Elsevier
41. Lee C-H, Oh J-E, Joe Y-G, Lee YY (2004) The performance improvement for an active noise control of an automotive intake system under rapidly accelerated conditions. JSME Int J Ser C 47(1):314–320
42. Beranek LL, Ver IL (1992) Noise and vibration control engineering-principles and applications. Noise and vibration control engineering-Principles and applications. Wiley, 814 p
43. Thilagavathi G, Pradeep E, Kannaian T, Sasikala L (2010) Development of natural fiber nonwovens for application as car interiors for noise control. J Ind Text 39(3):267–278
44. Parikh D, Chen Y, Sun L (2006) Reducing automotive interior noise with natural fiber nonwoven floor covering systems. Text Res J 76(11):813–820
45. Tascan M, Vaughn EA (2008) Effects of total surface area and fabric density on the acoustical behavior of needle punched nonwoven fabrics. Text Res J 78(4):289–296
46. Yilmaz ND (2009) Acoustic properties of biodegradable nonwovens
47. Arenas JP, Crocker MJ (2010) Recent trends in porous sound-absorbing materials. Sound Vib 44(7):12–18
48. Chen W-H, Lee F-C, Chiang D-M (2000) On the acoustic absorption of porous materials with different surface shapes and perforated plates. J Sound Vib 237(2):337–355
49. Lou C-W, Lin J-H, Su K-H (2005) Recycling polyester and polypropylene nonwoven selvages to produce functional sound absorption composites. Text Res J 75(5):390–394
50. Mohrova J, Kalinova K (2012) Different structures of PVA nanofibrous membrane for sound absorption application. Journal of Nanomaterials 2012:11
51. Li S, Wang Y, Ding J, Wu H, Fu Y (2014) Effect of shear thickening fluid on the sound insulation properties of textiles. Text Res J 84(9):897–902
52. Na Y, Agnhage T, Cho G (2012) Sound absorption of multiple layers of nanofiber webs and the comparison of measuring methods for sound absorption coefficients. Fibers and Polymers 13(10):1348–1352
53. Xu L, Liu F, Faraz N (2012) Theoretical model for the electrospinning nanoporous materials process. Comput Math Appl 64(5):1017–1021
54. Mahish S, Nayak R (2007) Coir fibre. Asian Textile J Bombay 16(10):54
55. Nayak R, Padhye R, Sinnappoo K, Arnold L, Behera BK (2013) Airbags. Text Progr 45
56. Patnaik A, Mvubu M, Muniyasamy S, Botha A, Anandjiwala RD (2015) Thermal and sound insulation materials from waste wool and recycled polyester fibers and their biodegradation studies. Energy Build 92:161–169
57. del Rey R, Alba J, Arenas JP, Sanchis VJ (2012) An empirical modelling of porous sound absorbing materials made of recycled foam. Appl Acoust 73(6):604–609
58. Nayak R, Padhye R (2015) Garment manufacturing technology. Elsevier
59. Seddeq HS, Aly NM, Elshakankery M (2012) Investigation on sound absorption properties for recycled fibrous materials. J Industrial Textiles:1528083712446956
60. Ashori A, Nourbakhsh A (2009) Characteristics of wood–fiber plastic composites made of recycled materials. Waste Manag 29(4):1291–1295
61. Hong Z, Bo L, Guangsu H, Jia H (2007) A novel composite sound absorber with recycled rubber particles. J Sound Vib 304(1):400–406
62. Lee Y, Joo C (2003) Sound absorption properties of recycled polyester fibrous assembly absorbers. AUTEX Res J 3(2):78–84
63. Maderuelo-Sanz R, Nadal-Gisbert AV, Crespo-Amorós JE, Parres-García F (2012) A novel sound absorber with recycled fibers coming from end of life tires (ELTs). Appl Acoust 73(4):402–408

64. Jordeva S, Tomovska E, Trajković D, Popeski-Dimovski R, Zafirova K (2015) Sound insulation properties of structure designed from apparel cutting waste. Paper presented at the AUTEX World Textile Conference, Bucharest, Romania, June 10–12
65. Ersoy S, Küçük H (2009) Investigation of industrial tea-leaf-fibre waste material for its sound absorption properties. Appl Acoust 70(1):215–220
66. Yang H-S, Kim D-J, Kim H-J (2003) Rice straw–wood particle composite for sound absorbing wooden construction materials. Bioresour Technol 86(2):117–121
67. Huda S, Yang Y (2009) Feather fiber reinforced light-weight composites with good acoustic properties. J Polym Environ 17(2):131–142
68. Zulkifli R, Nor M, Tahir M, Ismail A, Nuawi M (2008) Acoustic properties of multi-layer coir fibres sound absorption panel. J Appl Sci 8(20):3709–3714
69. Czigány T (2006) Special manufacturing and characteristics of basalt fiber reinforced hybrid polypropylene composites: mechanical properties and acoustic emission study. Compos Sci Technol 66(16):3210–3220
70. Mueller DH, Krobjilowski A (2003) New discovery in the properties of composites reinforced with natural fibers. J Ind Text 33(2):111–130

Basics of Acoustic Science

Vinod V. Kadam and Rajkishore Nayak

Abstract The act of hearing sound is an important communication and sensory medium with the surroundings. Sound is a wave that can be defined scientifically similar to the other mechanical waves in physics. The science of sound is acoustics that deals with the propagation of mechanical waves in various mediums such as solid, liquid and gases. As modern lifestyle includes the applications of acoustics in many fields, it is essential to understand the science of sound for people involved in acoustic applications. Various disciplines of acoustics include musical designs, vibration, noise pollution control, audiology, speech, vibration, underwater communication, audio signal processing, automotive acoustics, aeroacoustics, structural acoustics, bioacoustics, ultrasound, vibration control and environmental noise control. This chapter deals with the concept of sound waves and basic terminologies associated with acoustic science. Sound generation and propagation is covered in detail. Furthermore, sound measurement and different sound classification systems are also covered.

Keywords Sound wave · Propagation · Frequency · Absorption · Decibel

1 Introduction

The most important interactive communication in the social sense of human beings is to hear [3]. A sensation or a feeling that we hear is a sound. The science of sound is termed as acoustics. The word acoustics originated from the Greek word meaning "to hear" [4]. The American National Standard and the Acoustical Society of America have defined sound as "(a) Oscillation in pressure, stress, particle displacement, particle velocity, etc., propagated in a medium with internal forces (e.g.,

V.V. Kadam
Central Sheep and Wool Research Institute, Avikanagar, Rajasthan 304 501, India

V.V. Kadam · R. Nayak (✉)
School of Fashion and Textiles, RMIT, University, Melbourne, VIC 3056, Australia
e-mail: s3514621@student.rmit.edu.au; rajkishore.nayak@rmit.edu.au

R. Padhye and R. Nayak (eds.), *Acoustic Textiles*, Textile Science and Clothing Technology, DOI 10.1007/978-981-10-1476-5_2

elastic or viscous), or the superposition of such propagated oscillation. (b) Auditory sensation evoked by the oscillation described in (a)". Hence, acoustics is a study of sound generation and propagation in different media.

Sound originates when a material or object vibrates. These vibrations propagate in solid, liquid or gas medium in a wave form from the emitter to the receiver. Thus, a sound wave is the transfer of energy emitted by a source material or object into the medium as it travels. A sound wave is characterized by its frequency, wavelength and amplitude. The interaction of sound waves with the receiver's surface can alter the wave characteristics depending upon the surface properties of the receiver material or object. The sound wave can be absorbed, transmitted, reflected, refracted and diffracted from the surface. The change in sound characteristics due to the above phenomenon has been utilized in many scientific applications including acoustic textiles.

Acoustic textiles may be important to reduce noise pollution in modern days. It is important to understand the basics of sound in order to develop an efficient product design, say for sound insulation or so. The sound pressure levels, sound intensity and sound classification systems can help in assessing the performance of the acoustic structure. This chapter describes the concept of sound generation, sound wave propagation and the characteristic features. This also includes sound interference such as absorption, transmission, reflection, refraction and diffraction.

2 Sound Generation

Any source of vibration, which disturbs air molecules, creates sound. It pushes and pulls air molecules that convert the vibrations into acoustic signals, known as sound. Sound perceived depends on three things: vibration source to form a sound wave, wave carrier medium (such as air) and a receiver to detect the sound [5] (Fig. 1). Sound source oscillates and brings the surrounding air into motion and in the presence of a recipient, sound can be perceived [6]. Sound is a mechanical disturbance that travels through an elastic or viscous medium at a speed depending on the characteristic of that medium [7]. Sound is a wave motion in an elastic media such as air, water or a rock. While air tubulates, the mass and momentum sources of a material are ultimately sound generators [8].

Fig. 1 Sound generation

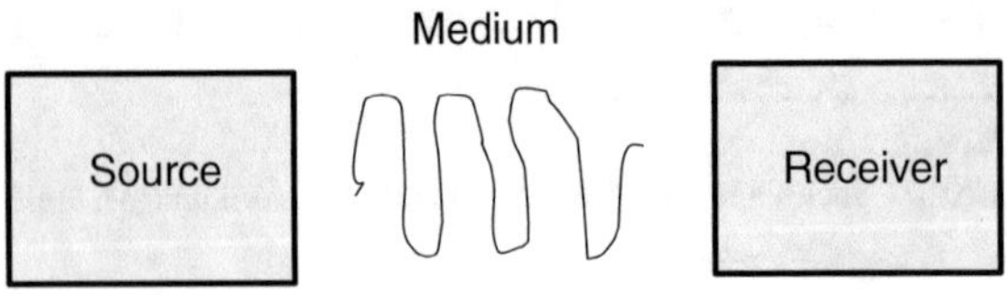

3 Sound Wave Propagation

Sound propagation is essentially a wave phenomenon. Acoustic signals require a mechanically elastic medium for propagation and therefore cannot travel through vacuum, unlike electromagnetic light waves. Sound travels more rapidly through solids, followed by liquids, than through gases [7]. Sound, in a wave form, travels at 331.29 ms^{-1} in dry air at a temperature of 0 °C [2].

3.1 Sound Wave Concept

When vibration sound disturbs particles in the air or the other medium, then those particles displace other surrounding particles. This particle movement goes on continuously in the outward direction to form a wave pattern. The wave carries the sound energy through the medium and becomes less intense as it moves away from the source. The sound energy is also directly associated with the volume of the sound. Higher sound energy results in loud volume.

There are three aspects of a sound wave that cause different types of sounds to be produced: frequency, wavelength and amplitude. Sound waves vibrate at different rates or frequencies as they move through the medium. The wave may have a single frequency or many frequencies depending upon the vibration source.

Let us consider a sound wave of constant frequency generated by a source with displacement function on Y-axis and time function on X-axis (Fig. 2). The number of waves generated per second is the frequency of the sound and expressed in hertz (Hz). The maximum displacement of a peak is termed as amplitude while the distance from one peak to the other is the wavelength. The sinusoidal wave generates only single frequency. A variety of non-repetitive sounds produce waves of different frequencies.

Sound waves are principally longitudinal waves. It means the wave medium, for instance, air, oscillates parallel to the wave's direction. Let us consider a soft coil is stretched and fixed at one end. If the coil is quickly pushed and pulled from the other

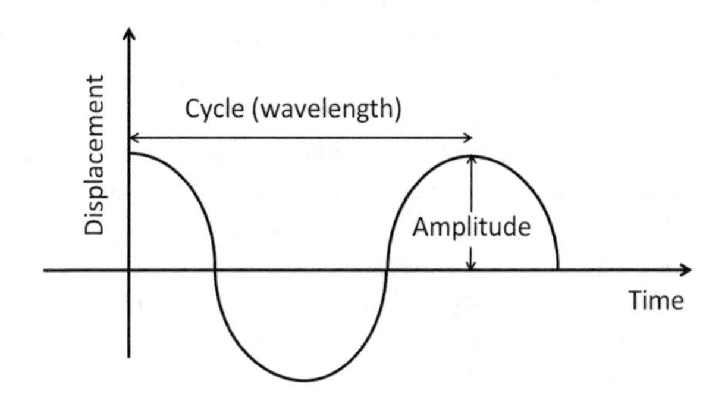

Fig. 2 Sound displacement-time diagram

end, it will compress and elongate along with the force direction. The same thing happens in longitudinal sound wave. Air particles get oscillated back and forth in the direction parallel to the sound wave movement. This create compression and rarefaction waves alternately. Longitudinal waves begin with compression followed by rarefaction. The wavelength can be determined by measuring the distance between two consecutive compressions, or rarefactions.

3.2 Sound Interference and Doppler Effect

When sound waves from two different sources either cancel or superimpose each other's effect, the phenomenon may be termed as interference [5]. The interference that cancels the waves generates no sound, whereas, noise or intermittent sound may be produced due to interference where the waves superimpose.

When a sound source is in motion, sound waves of variable lengths are generated. Wavelength become short in one direction and extends in the other. It causes change in the pitch of the sound perceived by a stationary receiver. This effect is commonly known as "Doppler effect" as per the name of Austrian Physicist Christian Doppler who discovered it [5]. In the Doppler effect, the resultant frequency reaching to the stationary receiver is not the actual output frequency of the sound source. The frequency is influenced by change in the wavelength depending on the sound source movement from the receiver. As the sound source moves towards a stationary receiver, wavelength is reduced causing an increase in the sound frequency and vice versa.

3.3 Absorption

The sound wave interact with the material or object surface and may be absorbed, transmitted, reflected, refracted or diffracted form the surface depending on type of the surface. These phenomenons are described in Fig. 3. When all the emitted sound waves are absorbed by the receiver, sound absorption occurs. It is exactly like sponge absorbing water. Sound absorption is an important phenomenon as far as sound insulation is concerned. There are different materials available for sound absorption. The sound absorbers may be porous or resonant type. Porous absorbents are classified as fibrous materials and open-celled foams. Fibrous materials convert acoustic energy into heat energy when sound waves impinge the absorber. In case of foam, sound wave displacement occurs through a narrow passage of foam and causes heat loss. Resonance absorbents are of mechanical type, where there is a solid plate with a tight air space behind. It is noteworthy that some material such as foam absorbs sound waves whereas the glass blocks it. The selection of material to be used depends on the end use application. For example, the office room in a building can be designed as sound absorbing or sound proofing.

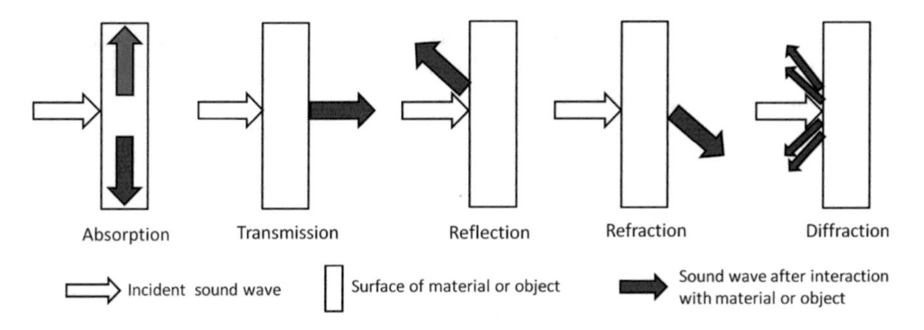

Fig. 3 Sound wave interaction with material or object surface

Sound absorption measures the amount of energy absorbed by the material and expressed as sound absorption coefficient (α). The coefficient ranges between 0 and 1 where 0 is no absorption and 1 is highest or total absorption. The higher coefficient yields lower reverberation time. The reverberation time is persistence of sound in a space after a sound source has been stopped [1]. It is the time lag, in seconds, for the sound to decay by 60 dB after a sound source has been stopped. Sound absorption is important to make the acoustic environment suitable for a specific purpose; for instance, in recording studios, lecture halls, concert rooms, lecture theatres, etc. The low frequency sound of 500 Hz is relatively difficult to absorb than high frequency sound.

3.4 *Transmission*

Sound waves from the source propagate through the medium and receiver without being absorbed or reflected and pass through the receiver without any frequency loss, which is known as sound transmission.

3.5 *Reflection*

When sound waves impinge on hard or smooth surface they may reflect back with their full energy without altering their characteristics. The reflection angle of sound wave from the reflecting surface is equal to the angle of incidence. The angles are defined between a normal to the reflecting plane and the incident and reflected waves. The reflected sound waves, thus, follow Huygen's geometry where both the incidence and reflection angles are equal [7].

The reflection phenomenon of sound waves finds many applications. For example, a reflected sound wave is used to measure the depth of water from sea level with the help of echo produced from the reflective surface. The geological composition at the bottom of the ocean and inside the earth crust is also identified using the reflection of sound wave [7]. Echo is a simple example of sound reflection phenomenon. Echo can be heard when the sound wave, perpendicular to the sound source, hits a flat and smooth surface.

3.6 *Refraction*

Refraction occurs when sound waves transmit through the surface and bent away from the straight line of travel. Sound refraction depends on factors such as the speed of sound, angle between sound propagation direction and wind direction and atmospheric conditions such as temperature and relative humidity [9].

3.7 *Diffraction*

Diffraction involves a change in the direction of sound waves as it strikes through a surface. Sound waves when impact on a partial barrier, some of them get reflected, some propagate without any disturbance and some bent or diffract over the top of the barrier. As sound source moves closer to the barrier, less sound diffraction is obtained. The sound at lower frequencies tends to diffract more easily than sound at higher frequencies [7].

4 Octave and 1/3 Octave Band

In acoustics, the sound frequencies are divided into ten standard octave bands. Each octave band has a centre frequency (f_c) and each centre frequency doubles the previous one (Table 1). The geometric mean of upper band limit (f_u) and lower

Table 1 Octave bands

Lower band limit (f_L) (Hz)	Centre frequency (f_c) (Hz)	Upper band limit (f_u) (Hz)
22.4 Hz	31.5 Hz	45 Hz
45	63	90
90	125	180
180	250	355
355	500	710
710	1 kHz	1.4 kHz
1.4 kHz	2	2.8
2.8	4	5.6
5.6	8	11.2
11.2	16	22.4

Source Raichel [7]

Table 2 Sound frequency categories

Sound	Frequency
Infrasound	<20 Hz
Audible sound	20 Hz–20 kHz
Ultrasound	>20 kHz
Hypersound	>1 GHz

Source Raichel [7]

band limit (f_L) denotes the centre frequency (f_c). The band width is the difference between the upper and lower band limits.

Each octave band is divided into three sub-bands to obtain one-third octave band. Each successive frequency is higher by a cube root of 2 than the previous one. The centre frequency for one-third octave band ranges from 20 Hz to 20 kHz.

5 Sound Attributes

There are two attributes of sound: loudness and tone [6]. The physical amount of loudness is sound pressure, whereas the tone is the sound frequency which is expressed in Hertz (Hz). The tone may also be termed as pitch of the sound. The pitch is also expressed as a wavelength which is the sound velocity per unit of sound frequency [5].

Human ear is not sensitive to all the sound frequencies. The sound may or may not be audible to the human ear, depending on its frequency and intensity. The human voice lies within 500–4000 Hz and it is most receptive to sounds within that range [10]. The audible sound for humans is in the range of 20 Hz–20 kHz [7]. Based on the frequency, sound is categorized as infrasound, audible sound, ultrasound and hypersound (Table 2).

6 Sound Pressure Level

Sound pressure level is a measure of volume (loudness) of the sound in terms of the sound pressure. The level can be determined by measuring the sound pressure disturbance from the equilibrium pressure value. The pressure disturbance is the difference between the instantaneous pressure and the static pressure [10]. The mean pressure deviation from the equilibrium is always zero, since the mean compression waves are equal to mean rarefaction waves. These positive and negative effects are converted into positive using the root mean square (RMS) value of sound pressure (P_{rms}) over a period of time. However, RMS value of sound pressure is not convenient to use as it varies over a wide range of magnitudes. The

Table 3 Sound pressure levels of different activities

Source	Sound pressure level (dB)
Aeroplane take off	140
Firecrackers	120
Rock concert	110
Factories	90
Traffic	80
Business places	60
Living room and kitchen	40
Bedroom	20
Natural places	10

Source Ziaran [11]; Department of Environment and Heritage, Queensland, Australia [12]

decibel (dB) is the easiest and a more convenient way to measure the volume (loudness) of the sound in terms of the sound pressure.

$$\mathrm{dB} = 20\log\left(\frac{p_{\mathrm{rms}}}{p_{\mathrm{o}}}\right)$$

dB Sound pressure level
rms Root mean square value
p_{o} Reference pressure.

Decibel is a value on a logarithmic scale and it is based on the capacity of humans to sense sound pressure. Sound perception by humans is subjective in nature. Different exposure times of the same sound pressure may have different effects on hearing [10]. In general, it is recommended that sound pressure levels should not exceed 30 and 40 dB in resting room and kitchen, respectively (ISO 25267:2004/T1:2009). The sound pressure level beyond 90 dB may be harmful for human hearing, especially when the exposure time is high. The sound pressure levels of day-to-day activities are shown in Table 3.

7 Sound Intensity

The sound intensity refers to the transfer of the sound wave energy that is a product of sound pressure and particle velocity

$$I = p \cdot u,$$

where I = sound intensity; p = sound pressure; u = particle velocity.

The sound intensity is associated with the sound power and surface area surrounding the source.

Table 4 Sound absorption class and absorption coefficient

Sound absorption class	Sound absorption coefficient
A	0.90; 0.95; 1.00
B	0.80; 0.85
C	0.60; 0.65; 0.70; 0.75;
D	0.30; 0.35; 0.40; 0.50; 0.55
E	0.15; 0.20; 0.25
F	0.00; 0.05; 0.10

Source ISO 11654:1997

8 Sound Insulation

Sound insulation is soundproofing of an enclosed space. It is apparent that when all the sound waves absorbed by the receiver, the enclosed space is said to be sound insulated. In acoustic science, sound insulation has become important particularly in modern days to prevent noise pollution. Sound insulation material is inserted in the place where sound insulation is required; for instance, in residential and business places. The insulating material such as fibre glass, foam, etc. can be imparted in ceiling, walls and bottom of the floor. The efficacy of the different sound absorbing materials can be evaluated using a grade scale. This is referred as sound classification system.

9 Sound Classification

In sound classification, as mentioned earlier, sound absorption by different sound insulation materials are represented by a common value, say, sound absorption coefficient. Based on the coefficient value, various sound absorption classes are developed. ISO 11654:1997 have standardized those classes as A–F where A is the best sound class with the highest sound absorption coefficient and F is the lowest sound absorption coefficient (Table 4).

10 Conclusion

Sound is a wave similar to other mechanical waves in physics. The science of sound that deals with the propagation of mechanical waves in various mediums such as solid, liquid and gaseous is known as acoustics. Sound originates when a material or an object vibrates. These vibrations propagate in solid, liquid or gas medium in a wave form from emitter to receiver. Thus, a sound wave is the transfer of energy emitted by a source material or an object into the medium as it travels. This chapter has discussed on the fundamentals of sound and its attributes, which will enable understanding of the other chapters of this book.

References

1. Reverberation time (2009). Oxford University Press
2. Speed of sound (2016). Britannica Academic
3. Blauert J (2005) Analysis and synthesis of auditory scenes. In: Communication acoustics. Springer
4. Blauert J, Xiang N (2009) Acoustics for engineers: troy lectures. Springer Science & Business Media
5. Hollar S (2013) Sound. Britannica Educational Publishing, Chicago
6. Möser M (2009) Engineering acoustics: an introduction to noise control. Springer Science & Business Media
7. Raichel DR (2006). The science and applications of acoustics. Springer Science & Business Media
8. Williams JF, Hawkings DL (1969) Sound generation by turbulence and surfaces in arbitrary motion. Philos Trans R Soc Lond A Math Phys Eng Sci 264:321–342
9. Wilson DK (2003) The sound-speed gradient and refraction in the near-ground atmosphere. J Acoust Soc Am 113:750–757
10. Wintzell L (2014) Acoustic textiles-the case of wall panels in home environment
11. Ziaran S (2005) Acoustics and acoustic measurements. Physical Methods, Instruments and Measurements. Vol- II, [Ed. Tsipenyuk YM], in Encyclopedia of Life Support Systems
12. Noise measurement manual (2013) Department of Environment and Heritage, The State of Queensland, Australia

Principles of Sound Absorbers

Xiaojun Qiu

Abstract This chapter introduces the principles of sound absorbers and the factors that affect acoustic absorption. The basic wave phenomena related to sound propagation such as acoustic reflection, absorption and scattering are introduced first and then the corresponding parameters such as transmission loss, absorption coefficient, scattering coefficient and flow resistance are explained. Sound absorbers can be made from porous materials or resonant structures, and the main mechanisms for sound absorption are acoustic impedance matching on the absorbers' boundary and acoustic energy dissipation inside the absorbers. Porous absorbers are materials where sound propagation occurs in a network of interconnected pores so that viscous and thermal effects cause acoustic energy to be dissipated. Resonant absorbers have two common forms: membrane/panel absorbers and Helmholtz absorbers. The main difference between porous absorbers and resonant absorbers is that the former is effective for broadband from mid to high frequency while the latter is usually only effective in a narrow tunable low frequency band. The applications of sound absorbers are focused on the discussion of reverberation time control and sound pressure level control in rooms. Finally, specific discussions are given to acoustic textiles and its recent developments. The contents of this chapter are focused on the fundamental aspects of sound absorption and absorbers to serve as an introduction to students and designers with basic knowledge in mathematics and physics.

Keywords Sound reflection · Sound absorption · Porous absorbers · Resonant absorbers · Reverberation time · Sound pressure level

---

X. Qiu (✉)
School of Electrical and Computer Engineering, RMIT University,
Melbourne, VIC 3001, Australia
e-mail: xiaojun.qiu@rmit.edu.au

R. Padhye and R. Nayak (eds.), *Acoustic Textiles*, Textile Science and Clothing Technology, DOI 10.1007/978-981-10-1476-5_3

1 Principles of Sound Absorbers

This chapter introduces the principles of sound absorbers and the factors that affect acoustic absorption. The basic wave phenomena related to sound propagation such as acoustic reflection, absorption and scattering are introduced first and then the corresponding parameters such as transmission loss, absorption coefficient, scattering coefficient and flow resistance are explained. The mechanisms of porous absorbers and resonance absorbers are discussed next as well as designing factors that affect acoustic absorption. The applications of sound absorbers are focused on the discussion of reverberation time control and sound pressure level control in rooms. Finally, specific discussions are given to acoustic textiles and it recent developments. The contents are focused more on the fundamental aspects of sound absorption and absorbers to serve as an introduction to students and designers with basic knowledge in mathematics and physics.

The approaches that are commonly used for sound control can be classified into the following three categories:

- control the sound radiation and generation from sound sources
- control the propagation and transmission of the generated sound
- control the sound pressure received by listeners.

The applications of the above approaches depend on the specific problems to be solved under specific situations. For example, if measures can be taken around the sound sources, then the mechanisms of the sound generation or their radiation pattern can be modified to reduce the sound power generated by the sound sources. This is the preferred method and should be applied first whenever it is feasible. If the sound is already generated and radiated into spaces, different measures can be implemented depending on the properties of environments. In free field, sound barriers can be used to reflect the sound or diffract the sound so that the sound received at some locations is reduced. In enclosures, in addition to sound barriers, sound absorption materials and structures can be applied in the space or on the boundaries to absorb sound and to reduce the reflection of the sound energy. With the sound absorption treatments in enclosures, the reverberant sound pressure as well as the reverberation time can be reduced. If the sound pressure is still too high after the treatments on sound sources and propagation paths, personal protection measures can be taken, which can be enclosures and ear mufflers.

Two main challenges in the applications of the above-mentioned passive sound control approaches are: low frequency sound control and ventilation requirements. Active noise control is a method of reducing existing noise by the introduction of controllable secondary sources to affect generation, radiation, transmission and reception of the original primary noise source. It can provide better solutions to low frequency noise problems than the current passive noise control methods when there are weight and volume constraints. It also provides an alternative noise control solution to applications where current passive noise control methods cannot be applied. The fundamental theories and methods of active noise control have become well established over the last 30 years; however, successful industry and civil applications of the technology are still limited in some specific cases such as headsets and earplugs, propeller aircrafts and cars [1].

1.1 Sound Propagation, Reflection and Absorption

Sound is a longitudinal mechanical wave, where the displacement of the medium at each point is normal to the local wavefront surface when the disturbance is travelling in a medium [2]. Sound speed is different in different medium, and that in air under normal atmospheric pressure at 20 °C is about 343 m/s. There is an energy loss when sound propagates in the medium, for example, in textiles; however, the attenuation of sound caused by air can usually be neglected in the low frequency range. For example, the sound pressure level attenuation is about 1 dB for a sound wave of 250 Hz travelling 1000 m; however, for a sound wave at 4000 Hz, the attenuation can be from 24 to 67 dB depending on the temperature and humidity of the air [3]. More about the measures for describing the sound propagation loss in textiles will be introduced in Sect. 1.1.1.

When sound encounters different medium or space, discontinuity in the propagation, reflection happens, where the incident wave arriving at the boundary interact with it to produce waves travelling away from the boundary [2]. The reflected wave or scattered waves follow certain rules. For example, for the specular reflection in which a plane incident wave is reflected by a uniform plane boundary, the normal wavenumber component of the incident field is reversed on reflection, and the wavenumber component parallel to the boundary is unaltered, so the angle of reflection is equal to the angle of incidence.

There is usually an energy loss when sound is reflected from a boundary, and the changes in amplitude and phase taken place during the reflection can be represented by the complex reflection factor (R) or sound pressure reflection coefficient as

$$R = \frac{p_{\mathrm{r}}}{p_{\mathrm{i}}} \qquad (1.1.1)$$

where p_{r} and p_{i} are complex amplitudes of the reflected and incident waves, respectively. The sound pressure reflection coefficient is a property of the boundary, whose magnitude and phase values depend on the frequency as well as the direction of the incident wave. More about the acoustic absorption and scattering on the boundary will be introduced in Sects. 1.1.2 and 1.1.3. Because the propagation and absorption properties of a textile depend on its flow resistance, the flow resistance will be introduced in Sect. 1.1.4.

1.1.1 Propagation

For the sound wave propagating in porous textiles, the porous gas-filled medium is often treated as an equivalent uniform medium for analysis purpose; so a propagation factor $e^{j\omega t-\gamma x}$ is used to describe the dependence of the propagating wave on time t and the propagation coordinate x [3]. Here, $\omega = 2\pi f$ is the angular frequency and f is the wave frequency. The propagation constant γ is also called the propagation coefficient, which is a complex number and can be described by,

$$\gamma = \alpha + jk \tag{1.1.2}$$

where the propagation wavenumber $k = \omega/c$ is also called the phase coefficient, α is called the attenuation coefficient, and c is the speed of sound in the medium. The attenuation coefficient α describes the reduction in amplitude of a progressive wave with the distance in the propagation direction by $e^{-\alpha x}$. This coefficient is actually the amplitude attenuation coefficient instead of the energy attenuation coefficient.

When a sound wave is incident upon a partition or a textile layer, some of it will be reflected back to the incidence medium and some will be transmitted through the partition or the textile layer. The fraction of incident energy that is transmitted is called the transmission coefficient (τ). If there is no other energy dissipation or reflections, the relationship between the transmission coefficient and the attenuation coefficient is

$$\tau = 1 - \alpha^2 \tag{1.1.3}$$

The transmission loss, TL is defined as the logarithm of the reciprocal of the transmission coefficient as

$$TL = -10\log_{10}\tau \tag{1.1.4}$$

In general, the transmission loss through a porous layer depends upon the angle of incidence and is a function of material density, thickness, flow resistivity and frequency. In the low frequency range, the transmission loss of common porous layers is usually less than 20 dB, but can be greater than 20 dB in high frequency range. The transmission loss usually increases with the material density, thickness, flow resistivity and frequency.

1.1.2 Absorption

There is usually an energy loss when sound propagates in a medium. Although the energy loss can be caused by some kinds of energy dissipation such as absorption, the term "attenuation coefficient" is used in Sect. 1.1.1 to describe the reduction in amplitude of a progressive wave with the distance in the propagation direction in the medium. When the sound wave encounters different medium or space discontinuity in the propagation, reflection happens. The energy loss associated with the reflection from a boundary is described by the sound pressure reflection coefficient, which is the change in amplitude and phase of the propagating wave taken place during the reflection. It is a property of the boundary, and the value depends on the frequency as well as the direction of the incident wave.

Acoustic absorption coefficient at a boundary, also called sound absorption factor or sound power absorption coefficient, is defined as the fraction of the incident acoustic power arriving at the boundary that is not reflected, and is therefore regraded as being absorbed by the boundary [2],

$$\alpha = 1 - \frac{E_\mathrm{r}}{E_\mathrm{i}} \tag{1.1.5}$$

where E_r and E_i are the reflected acoustic power and the incident acoustic power arriving at the boundary, respectively. The values of the acoustic absorption coefficient can range from 0 to 1.0. A value of 0 refers to the condition when there is a total reflection, whereas 1.0 indicates perfect absorption and no reflection. It is also a function of frequency and incident wave direction. The (random incidence) statistical absorption coefficient can be obtained by,

$$\alpha_\mathrm{st} = \frac{1}{\pi}\int_0^{2\pi} \mathrm{d}\varphi \int_0^{\pi/2} \alpha(\theta)\cos\theta\,\sin\theta\,\mathrm{d}\theta \tag{1.1.6}$$

where θ and φ are the angle of elevation and azimuth, respectively. The Sabine absorption coefficient is the random incidence absorption coefficient deduced from the reverberation time measurement via the Sabine equation.

Porous absorbers are materials where sound propagation occurs in a network of interconnected pores (open pore structure) in such a way that viscous and thermal effects cause acoustic energy to be dissipated. Air is a viscous fluid, and consequently acoustic energy is dissipated via friction with the pore walls. As well as viscous effects, there will be losses due to the thermal conduction from the air to the absorber material, which might be more significant at low frequency. For a porous absorber to create significant absorption, it needs to be placed somewhere where the particle velocity is high. The particle velocity close to a room boundary is usually zero, so the parts furthest from the backing surface are often most effective. The material needs to be greater than one-tenth wavelength thick to cause significant absorption, and about one-fourth wavelength to absorb most incident sound. More will be discussed in Sect. 1.2.

It is often difficult to obtain absorption at low frequencies with porous textile absorbers because the required thickness of the material is large and the treatments are often placed at room boundaries where the absorbers are inefficient due to the low particle velocity. Resonant absorbers can be an alternative solution. There are two common forms of resonant absorbers: membrane/panel absorber and Helmholtz absorber. For a membrane/panel absorber, the mass is usually a sheet of mass-loaded vinyl or plywood which vibrates; while for a Helmholtz absorber, the mass is a plug of air in the opening of the perforated sheet. The spring in both the cases is provided by air enclosed in the cavity. The resonant frequency of this type of absorbers can be tuned. More will be discussed in Sect. 1.3. Sound absorbers are usually employed for adjusting the reverberation of the room, suppressing undesired sound reflections from remote walls (echoes), and reducing the acoustical energy density and hence the sound pressure level in noisy rooms. More will be discussed in Sect. 1.5.

1.1.3 Scattering

The degree of scattering and absorption in a room are important factors related to the acoustic quality of the room. Although scattered polar responses can give much information about the scattering from a textile surface, it yields too much detail, and a single value is desired in practice to allow easy comparison of diffuser quality. There is also a need for a scattering coefficient to evaluate the amount of dispersion generated by a surface to allow accurate predictions using geometric room acoustic models [4]. The geometric models use a scattering coefficient to determine the proportion of the reflected energy that is reflected in a specular manner and the proportion that is scattered.

The scattering coefficient, s_θ for an incident wave arriving at the boundary with an angle θ is defined as the value calculated by one minus the ratio of the specularly reflected acoustic energy to the total reflected acoustic energy,

$$s_\theta = 1 - \frac{E_\mathrm{s}}{E_\mathrm{r}} \tag{1.1.7}$$

where E_s is the specularly reflected acoustic energy and E_r is the total reflected acoustic energy from the incident acoustic power arriving at the boundary. Theoretically, s_θ takes a value between 0 and 1, where 0 means a totally specularly reflecting surface and 1 means a totally scattering surface. When there is no subscript, random incidence scattering coefficient s is used, which is defined as the value calculated by one minus the ratio of the specularly reflected acoustic energy to the total acoustic energy reflected from a surface in a diffuse sound field.

The directional diffusion coefficient $d_{\theta,\varphi}$ is a measure of the uniformity of diffusion produced by a surface for one sound source [4],

$$d_{\theta,\varphi} = \frac{\left(\sum_{i=1}^{n} 10^{L_i/10}\right)^2 - \sum_{i=1}^{n} \left(10^{L_i/10}\right)^2}{(n-1)\sum_{i=1}^{n} \left(10^{L_i/10}\right)^2} \tag{1.1.8}$$

where the subscript θ is the angle of incidence relative to the reference normal of the surface, and φ indicates the azimuth angle, n is number of receivers, L_i are a set of sound pressure levels in a polar response of the scattering sound. The directional

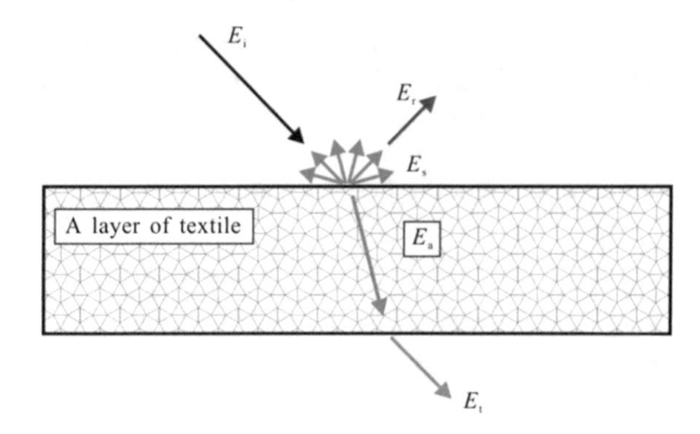

Fig. 1 Energy relationships when a propagating wave is incident upon a lay of textile

diffusion coefficient has a value between 0 and 1, corresponding to that one receiver receives non-zero scattered sound pressure or complete diffusion, respectively. The random incidence diffusion coefficient calculated from weighting the directional diffusion coefficients for the difference source positions is a measure of the uniformity of diffusion for a representative sample of sources over a complete semicircle for a single plane diffuser, or a complete hemisphere for a hemispherical diffuser.

Both the scattering and diffusion coefficients are simplified representations of the true reflection behaviour [4]. The purpose of the diffusion coefficient is to enable the design of diffusers and to allow acousticians to compare the performance of surfaces for room design and performance specifications. While the scattering coefficient is to characterize surface scattering for use in geometrical room modelling programs, the diffusion coefficient is different from, but related to, the random incidence scattering coefficient. The scattering coefficient is a rough measure of the degree of scattered sound, while the diffusion coefficient describes the directional uniformity of the scattering, i.e. the quality of the diffusing surface.

Figure 1 summarizes the relationships of different kinds of energy when a propagating wave is incident upon a layer of textile. The total input energy brought from the incident wave is E_i, which is equal to the summation of E_r, E_s, E_a and E_t, i.e., the energy reflected and scattered from the boundary, the energy dissipated inside the textile layer and the energy transmitted through the layer.

1.1.4 Flow Resistivity

Most textiles that can be used for acoustic purposes show open porosity due to many interconnected pores or voids inside. The acoustic performance of a porous textile is mainly determined by its flow resistivity, which is an intrinsic property of the textile and is a measure of how easily air can enter a porous textile material and the resistance that the air flow meets through the material [4]. Flow resistivity is also known as static flow resistivity and is used to describe some of the structural properties in an indirect manner. It may be used to establish correlations between the structure of the materials and some of the acoustical properties. It plays a critical role in the calculation of many intrinsic acoustic properties of porous textiles, such as the characteristic impedance, the propagation constant and the sound absorption coefficient. Flow resistivity has a unit of N.s.m^{-4} and is defined as the unit-thickness specific flow resistance (σ) by [2],

$$\sigma = \frac{R_\mathrm{f}}{h} \tag{1.1.9}$$

where R_f is the specific flow resistance of a uniform textile layer of thickness h and is defined as

$$R_\mathrm{f} = \frac{\Delta p\, S}{V_0} \tag{1.1.10}$$

where Δp is the pressure drop through the layer with a surface area of S under conditions of slow steady flow with a volume flow rate of V_0.

(Static) flow resistance is the ratio of the pressure drop across a porous element to the volume velocity flowing through it under conditions of steady low speed flow. The flow resistance is almost independent of the volume velocity at low speeds; however, it is frequency dependent. Dynamic specific flow resistance of a thin (compared to an acoustic wavelength) porous textile layer is the real part of the complex specific flow impedance at a specified frequency, which is defined as the complex ratio of the pressure drop across the layer to the relative face velocity through the layer. When the frequency approaches zero, the dynamic specific flow resistance varies little with frequency, so it almost equals to the static flow resistance. For fibrous porous materials such as fibreglass and rockwool, their flow resistivity can be estimated with [3]:

$$\sigma = 27.3 \left(\frac{\rho_\mathrm{m}}{\rho_\mathrm{f}}\right)^{k_1} \frac{\mu}{d^2} \tag{1.1.11}$$

where ρ_m is the porous material bulk density, ρ_f is the fibre material density, μ is the dynamic gas viscosity (1.84×10^{-5} kg m^{-1} s^{-1} for air at 20 °C), d is the fibre diameter of the porous material, and the constant $k_1 = 1.53$ for fibreglass and rockwool with a diameter between 1 and 15 μm and might be a different value for other fibres and different diameters.

1.2 Porous Absorbers

Porous absorbers are materials where sound propagation occurs in a network of interconnected pores in such a way that viscous and thermal effects cause acoustic energy to be dissipated [4]. Carpets, acoustic tiles, open cell acoustic foams,

Fig. 2 Illustration of an open pore structure for sound absorption

curtains, cushions, cotton and mineral wools such as fibreglass are such materials. Air is a viscous fluid, and consequently acoustic energy is dissipated via friction with the pore walls provided that the size of pore is sufficiently small. As well as viscous effects, there are losses due to thermal conduction from the air to the absorber material, especially at low frequency. For the absorption to be effective there must be interconnected air paths through the material; so an open pore structure is necessary, as shown in Fig. 2 [5]. The viscous and thermal effects are affected by the pore diameter, network shape and layout, density and other physical properties of the material such as cell size and thickness, so the micromodels for understanding the underneath physical mechanisms of sound absorption in porous materials are quite complicated.

Fortunately, these micro effects for sound absorption of a uniform textile layer can be approximately represented by a macro quantity, i.e. the specific flow resistance. As introduced in Sect. 1.1.4, the flow resistivity of a textile layer is an intrinsic property of the textile and is a measure of how easily air can enter a porous textile material and the resistance that the air flow meets through the material. Equation (1.1.11) shows that the flow resistivity of certain fibreglass and rockwool is proportional to the dynamic gas viscosity and the ratio of the porous material bulk density to the fibre material density, but inversely proportional to the squared fibre diameter; however, it varies little with frequency.

For the sound wave propagating in porous textiles, the porous gas-filled medium is often treated as an equivalent uniform medium for analysis purpose, which can be characterized in dimensionless variables by a complex density and complex compressibility [3]. These quantities can be calculated from the flow resistivity and then be used to calculate the characteristic impedance and the complex propagation constant of the porous material. The following empirical formulas can be used to estimate the characteristic impedance, Z_m, and the complex propagation constant, k_m, of the porous material directly from the flow resistivity for fibreglass and rockwool materials with a small amount of binder and having short fibres smaller than 15 μm in diameter [6],

$$Z_{\mathrm{m}} = \rho_0 c_0 \left[1 + 0.0571 \left(\frac{\rho_0 f}{\sigma} \right)^{-0.754} - \mathrm{j}0.087 \left(\frac{\rho_0 f}{\sigma} \right)^{-0.732} \right] \quad (1.2.1)$$

$$k_{\mathrm{m}} = k_0 \left[1 + 0.0978 \left(\frac{\rho_0 f}{\sigma} \right)^{-0.700} - \mathrm{j}0.189 \left(\frac{\rho_0 f}{\sigma} \right)^{-0.595} \right] \quad (1.2.2)$$

where ρ_0 is the air density, c_0 is the speed of sound, $k_0 = 2\pi f/c_0$ is the wavenumber in air and f is the wave frequency.

When a sound wave is incident upon a textile layer, some of the energy is reflected back to the incidence medium and some will be transmitted through the partition or the textile layer. The absorption coefficient at the boundary is the fraction of the incident acoustic power arriving at the boundary that is not reflected, and its value depends not only on the properties of the material but also on the properties of the space behind the material. For a layer of textile, its absorption coefficient when it is backed by a rigid wall is different to that when it is hanged in a room. The value depends on the distance between the layer and the walls as well as the angle of incidence. For the simplest case of normal incidence with a plane wave, its absorption coefficient can be calculated with the reflection coefficient (R) as follows [3],

$$\alpha = 1 - |R|^2 = 1 - \left| \frac{Z_{\mathrm{s}} - \rho_0 c_0}{Z_{\mathrm{s}} + \rho_0 c_0} \right|^2 \quad (1.2.3)$$

where the specific acoustic impedance at the layer's front surface, Z_{s}, can be formulated with the impedance transfer function for a layer of porous material with a thickness of l by [7] and, [8],

$$Z_{\mathrm{s}} = Z_{\mathrm{m}} \frac{Z_{\mathrm{L}} + \mathrm{j} Z_{\mathrm{m}} \tan(k_{\mathrm{m}} l)}{Z_{\mathrm{m}} + \mathrm{j} Z_{\mathrm{L}} \tan(k_{\mathrm{m}} l)} \quad (1.2.4)$$

Z_{m}, k_{m} and Z_{L} are respectively the characteristic impedance, propagation constant and the acoustic impedance at the back surface of the porous material. For the configuration shown in Fig. 3, when the wall is rigid and the distance between the layer and the wall is L, Z_{L} can be calculated by

$$Z_{\mathrm{L}} = -\mathrm{j} \rho_0 c_0 \cot(k_0 L) \quad (1.2.5)$$

where $k_0 = \omega/c_0$ is the wavenumber in the air.

For a layer of textile placed against the rigid wall, $L = 0$, so $Z_{\mathrm{L}} \to -\infty$, Eq. (1.2.4) becomes,

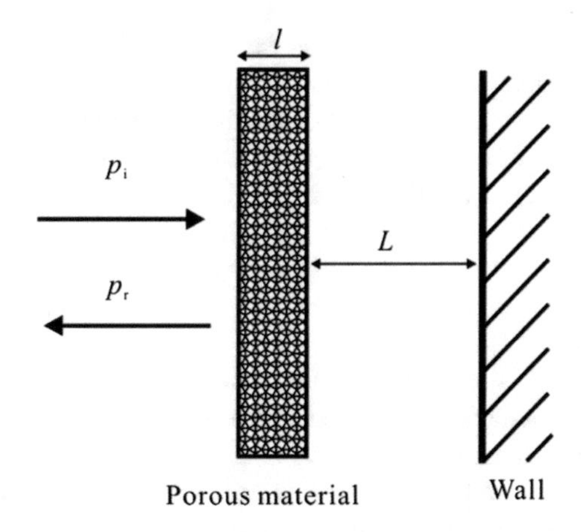

Fig. 3 A layer of porous material with a thickness of l located with a distance of L from a wall

$$Z_s = -jZ_m \cot(k_m l) \tag{1.2.6}$$

The characteristic impedance in Eq. (1.2.1) can be denoted simply as $Z_m = \rho_p c_p$, where ρ_p is the complex density and c_p is the speed of sound in the layer of the porous material. The complex propagation constant in Eq. (1.2.2) can also be denoted simply as $k_m = k_p - j\alpha_p$, where the propagation wavenumber $k_p = 2\pi f/c_p$ is the phase coefficient and α_p is called the attenuation coefficient in the layer of the porous material. Substitute them into Eq. (1.2.6), it has

$$Z_s \approx -j\rho_p c_p \cot(k_p l - j\alpha_p l) \tag{1.2.7}$$

Normalizing this specific acoustic impedance at the layer's front surface with the characteristic impedance of air yields

$$\frac{Z_s}{\rho_0 c_0} = \frac{-j\rho_p c_p \cot(k_p l - j\alpha_p l)}{\rho_0 c_0} = \gamma A + j\gamma B \tag{1.2.8}$$

where $\gamma = \rho_p c_p / \rho_0 c_0$ is the ratio of the characteristic impedance of the porous material to that of the air and

$$A = \frac{\sinh(2\alpha_p l)}{\cosh(2\alpha_p l) - \cos(2k_p l)}, \quad B = \frac{-\sin(2k_p l)}{\cosh(2\alpha_p l) - \cos(2k_p l)} \tag{1.2.9}$$

Using Eq. (1.2.3), the absorption coefficient for a layer of porous material placed directly against the rigid wall under normal incidence is

$$\alpha = \frac{4\gamma A}{(1 + \gamma A)^2 + (\gamma B)^2} \tag{1.2.10}$$

This equation can be used to illustrate the sound absorption mechanisms of porous material. For example, when the sound attenuation capability of the material is sufficient strong or the thickness of the absorption material is sufficient long, i.e. $\alpha_p l \gg 1$, then $\sinh(2\alpha_p l)$ and $\cosh(2\alpha_p l)$ tend to infinite large, so $A \approx 1$ and $B \approx 0$, the absorption coefficient becomes

$$\alpha = \frac{4\gamma}{(1+\gamma)^2} \quad (1.2.11)$$

It is clear that the absorption coefficient can be 1 only when $\gamma = 1$, i.e. the characteristic impedance of the porous material equals to that in the air. The sound attenuation capability of a porous material is not equivalent to its sound absorption capability. To have perfect sound absorption capability, the characteristic impedance of the porous material needs to match that in the air so that no sound is reflected back first, and then the transmitted sound into the material can be absorbed or dissipated partially or completely by the material.

If $k_p l = \pi/2$, i.e. $l = \lambda_p/4$, λ_p is the wavelength of the sound in the material, then $A = \tanh(\alpha_p l)$ and $B = 0$, the absorption coefficient researches a maximal value,

$$\alpha_{\max} = \frac{4\gamma \tanh(\alpha_p l)}{[1+\gamma \tanh(\alpha_p l)]^2} \quad (1.2.12)$$

If $k_p l = \pi$, i.e. $l = \lambda_p/2$, then $A = 1/\tanh(\alpha_p l)$ and $B = 0$, the absorption coefficient researches a minimal value,

$$\alpha_{\min} = \frac{4\gamma \tanh(\alpha_p l)}{[\gamma + \tanh(\alpha_p l)]^2} \quad (1.2.13)$$

The corresponding frequencies to the maximal and minimal values are called the resonant frequency and anti-resonant frequency.

When there is a space between the layer of the material and the wall, i.e. $L \neq 0$, the variation of the absorption coefficient can be analyzed similarly with Eqs. (1.2.3)–(1.2.5). A general conclusion is that the space between the layer of the material and the wall increases its absorption value at low frequency. The physical explanation is that the particle velocity of sound wave close to rigid walls is usually zero, so placing the material somewhere further from the backing surface will have more absorption because of high particle velocity of sound wave.

Figure 4 shows typical sound absorption coefficient curves of a porous material as a function of frequency, where the solid line corresponds to that of a 5 cm thick porous material with a flow resistivity of 10,000 N s m^{-4} placed directly against a rigid wall, the dot line corresponds to that of the 5 cm thick porous material placed 10 cm away from the rigid wall, and the dash-dot line corresponds to that of the same porous material but with a thickness of 15 cm placed directly against a rigid wall. It is obvious that the absorption coefficient increases with frequency. The

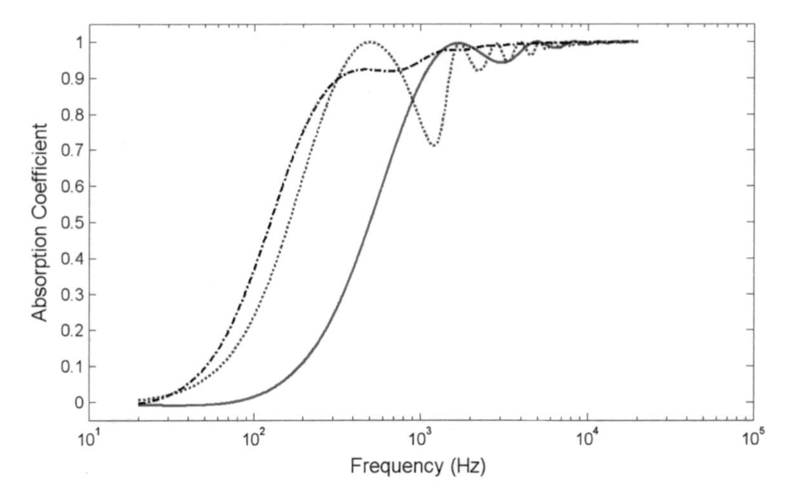

Fig. 4 Typical sound absorption coefficient curves of a layer of porous material with a flow resistivity of 10,000 N s m^{-4} as a function of frequency, *solid line* corresponds to that of a 5 cm thick material placed directly against a rigid wall, the *dot line* corresponds to that of the 5 cm thick porous material placed at 10 cm away from the rigid wall, and the *dash-dot line* corresponds to that of the same porous material but with a thickness of 15 cm placed directly against a rigid wall

mechanism for the low absorption coefficient at low frequency can be illustrated with Eq. (1.2.4). At very low frequency, $k_{\mathrm{m}}l \ll 1$, then $\tan(k_{\mathrm{m}}l)$ tends to be zero, thus Z_{s} is almost equal to Z_{L} and the porous material seems not existing. Z_{L} is the acoustic impedance at the back surface of the porous material, which can be calculated with Eq. (1.2.5) and is usually much larger than the characteristic impedance of the air, so most the incident acoustic energy will be reflected back because of impedance mismatching.

1.3 Resonance Absorbers

As shown in Sect. 1.2, it is often difficult to have large sound absorption at low to mid frequencies with porous absorbers in practice because the required thickness of the material is large and the treatments are often placed at room boundaries where porous absorbers are inefficient due to the low particle velocity. A different kind of sound absorber called resonant absorber can be used to solve the problem, which can tune its resonant frequency to a specified frequency and have maximal sound absorption there [4]. The resonant absorbers work on the principle of dissipating acoustic energy with structure vibration. The most significant difference between resonance absorbers and porous absorbers is that the former is usually only effective in a narrow tunable low frequency band while the latter is effective for broadband from mid to high frequency.

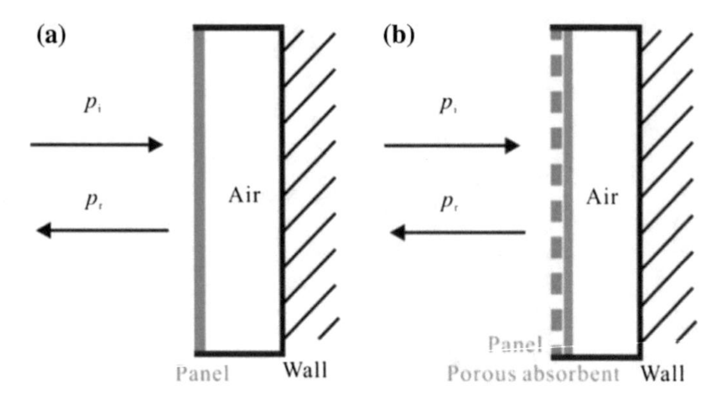

Fig. 5 A schematic diagram of: (**a**) a panel absorber and (**b**) a Helmholtz absorber

There are two common forms of the resonant absorbers: membrane/panel absorber and Helmholtz absorber. For a membrane/panel absorber, the mass is a vibrating sheet of membrane or panel made from various materials, the spring is usually provided by the resilient boundary of the membrane or panel or air enclosed in the cavity; while in the case of a Helmholtz absorber, the mass is a plug of air in the opening of a perforated sheet and the spring is usually provided by air enclosed in the cavity. It is often best to place some porous absorbent in the neck of a Helmholtz resonator or just behind the membrane or panel in a membrane/panel absorber to increase the acoustic resistance of the whole absorber. Figure 5 shows a schematic diagram of a panel absorber and a Helmholtz absorber.

The mechanism of sound absorption for membrane or panel absorbers is energy dissipation by vibration of the membrane or panel. Whether the flexible membranes or panels are mounted over an air space or are mounted on a suspended ceiling, the membranes or panels must couple with and be driven by the sound field. Acoustic energy is then dissipated by flexure of the membrane or panel. Additionally, if the backing air space is filled with a porous material, energy is also dissipated in the porous material. Maximum absorption occurs at the first resonance of the coupled membrane/panel-cavity system [3].

It should be noted that this mechanism of resonant absorbers is different to that of porous absorbers. For the porous absorbers, the absorption depends only upon local properties of the material, while for the membrane/panel absorbers the absorption is dependent upon the response of the panel as a whole. Furthermore, as the panel absorber depends upon strong coupling with the sound field to be effective, the energy dissipated is very much dependent on the sound field and thus on the rest of the room in which the panel absorber is used. This latter fact makes prediction of the absorptive properties of membrane/panel absorbers difficult. Two design methods (one is empirical based upon data measured in auditoria and concert halls and the other is based upon analysis) can be found in the Ref. [3] for estimating the Sabine absorption of panel absorbers. The two main steps in these methods are tuning the resonant frequency and adjusting the sound absorption peak value and band width; however, they will not be introduced here because they are quite complicated.

1.3.1 Helmholtz Resonant Absorber

The sound absorption mechanism of Helmholtz resonators is to form an acoustic resonant system so that the acoustic energy can be better dissipated with the acoustic resistance, thanks to the amplified particle velocity. The acoustic resistance can be obtained by having some porous material near the neck where the particle velocity is the maximal or to make the opening very small to have high acoustic resistance. Figure 6 shows a schematic diagram of a single Helmholtz absorber, which is an individual cell of Fig. 5b. The Helmholtz resonator consists of an air cavity with a volume of V acting like spring in a mechanical resonant system, a neck with a length of l_0 and a diameter of d. The mass of the air in the neck acts like the mass in the mechanical resonant system. Because there is no physical subdividing in the volume in Fig. 5b, the model in Fig. 6 is an approximation of that in Fig. 5b. However, when a porous absorbent is placed in the cavity, the need for physical subdividing is less critical [4].

For a single Helmholtz absorber shown in Fig. 6, the cavity walls are usually rigid, the volume of neck is smaller than that of the volume, and the dimension of the device is usually much smaller than the wavelength because the resonant frequency is quite low. The acoustic impedance of the absorber can be expressed as

$$Z_a = R_a + jX_a \tag{1.3.1}$$

where R_a is the acoustic resistance of the absorber, which is related to the porous material used in the cavity and/or the viscous loss along the neck. X_a is the acoustic reactance of the absorber and can be calculated from the mechanical reactance of the system by

$$X_a = \left(\omega M_a - \frac{K_a}{\omega}\right) \tag{1.3.2}$$

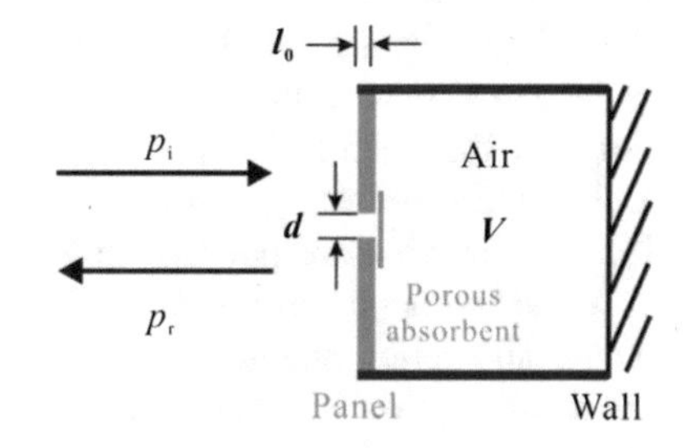

Fig. 6 A schematic diagram of a single Helmholtz absorber

where $\omega = 2\pi f$ is the angular frequency and f is the frequency, $M_a = \rho_0 l_e/S_0$ is the mass of the air in the neck and $l_e = l_0 + 0.8d$ is the equivalent length of the neck, $S_0 = \pi d^2/4$ is the cross-section area of the neck, $K_a = \rho_0 c_0^2/V$ is the acoustic stiffness of the air cavity as a spring. The local specific acoustic impedance of the absorber, $Z_s = R_s + jX_s$, can be obtained by

$$Z_s = Z_a S \tag{1.3.3}$$

where S is the cross-section area of the whole absorber to the incident sound. Similar to that for porous absorbers, the absorption coefficient can be calculated with the reflection coefficient as follows [3],

$$\begin{aligned} \alpha &= 1 - \left| \frac{Z_s - \rho_0 c_0}{Z_s + \rho_0 c_0} \right|^2 \\ &= \frac{4R_s/\rho_0 c_0}{(R_s/\rho_0 c_0 + 1)^2 + (X_s/\rho_0 c_0)^2} \end{aligned} \tag{1.3.4}$$

If $X_a = 0$, the absorption coefficient will be maximum, and the resonant frequency can be calculated by setting Eq. (1.3.2) to zero. The solved resonant frequency depends on the geometry of the Helmholtz absorber as shown in Eq. (1.3.5).

$$f_r = \frac{1}{2\pi}\sqrt{\frac{K_a}{M_a}} = \frac{c_0}{2\pi}\sqrt{\frac{S_0}{VL_e}} \tag{1.3.5}$$

To have low resonance frequency, the volume of the cavity should be large, the length of the neck should be long and the cross-section area of the neck should be small. Even at the resonant frequency, the acoustic resistance of the absorber needs to match the characteristic impedance of the air so that $R_s = \rho_0 c_0$ to have perfect absorption.

Unlike porous absorbers, resonant absorbers sometimes are not distributed uniformly against the boundaries but used as a single separate object. Under this situation, absorption coefficient which refers to a uniform surface might not be appropriate for describing their sound absorption performance. Another terminology, absorption cross section (or equivalent absorption area), which is defined as the ratio of sound power being absorbed and the intensity which the incident sound wave would have at the place of the absorber if it were not present, can be used to describe the sound absorption performance of individual absorbers [5]. For a Helmholtz absorber, its absorption cross section is approximately $\lambda^2/2\pi$, where λ is the wavelength corresponding to the resonant frequency.

If the Helmholtz absorbers are distributed uniformly as that shown in Fig. 5b, its sound absorption coefficient can be calculated with Eq. (1.3.4). Figure 7 shows sound absorption coefficient curves of a typical perforated sheet (a 6.3 mm thick panel with a perforation rate of 6 %, the diameter of the holes is 5 mm, and the air

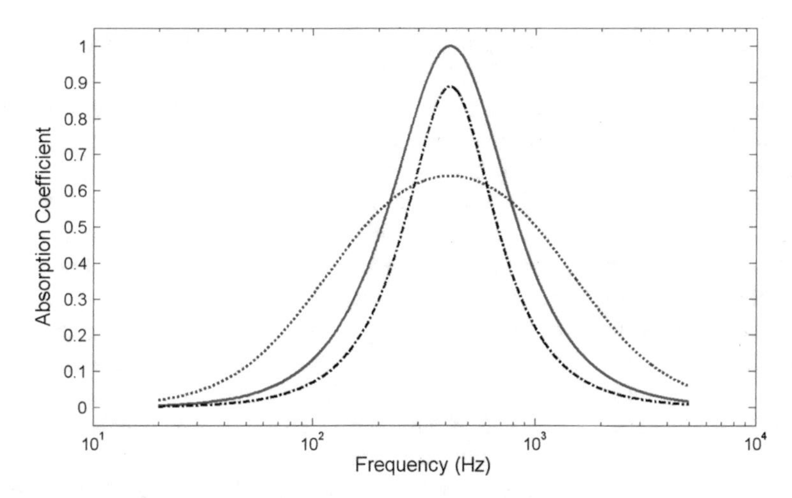

Fig. 7 Sound absorption coefficient curves of a typical perforated sheet (a 6.3 mm thick panel with a perforation rate of 6 %, the diameter of the holes is 5 mm, the air layer thickness between the panel and the wall is 0.1 m) as a function of frequency, where the *solid line* corresponds to that with $R_s = \rho_0 c_0$, the *dot line* corresponds to that with $R_s = 4\rho_0 c_0$, and the *dash-dot line* corresponds to that with $R_s = 0.5\rho_0 c_0$, the resonant frequency of the absorber is approximately 418 Hz

layer thickness between the panel and the wall is 0.1 m) as a function of frequency, where the solid line corresponds to that with $R_s = \rho_0 c_0$, the dot line corresponds to that with $R_s = 4\rho_0 c_0$, and the dash-dot line corresponds to that with $R_s = 0.5\rho_0 c_0$. It is clear that the acoustic resistance of the absorber needs to be finely tuned to have perfect absorption. The resonant frequency of the absorber is approximately 418 Hz under the current configuration, but can be reduced by increasing the air layer thickness between the panel and the wall or the length of the neck. It can also be reduced by reducing the size of the neck cross-section area as shown in Eq. (1.3.5).

1.3.2 Micro-Perforated Panel Resonators

A micro-perforated panel (MPP) consists of a thin sheet panel perforated with a lattice of sub-millimetre apertures which create high acoustic resistance and low acoustic mass reactance necessary for broadband sound absorption without further using additional porous material [9]. Because the light and fibreless MPPs can be made of various recyclable materials, they are becoming more and more widely used in sound field control today, especially in clean situations where strict hygiene is required or in industrial environments where porous materials deteriorate. Furthermore, MPPs can be made of transparent or colourful plates or membranes, so they are also in demand by architects for sound quality control in auditoriums [10]. MPP absorbers are tagged as the "next generation" absorbing materials due to their huge potential in comparison with conventional porous materials [11].

The mechanism of MPP absorption is related to the resonance effect, where the air inside the apertures of the MPP vibrates like a mass and the air inside the backing cavity acts like a spring, so the effective sound absorption frequency band is around the resonance peak. Overcoming this drawback (narrow absorption frequency range) has been the subject of a significant amount of recent work [12–21]. For example, additional MPPs can be used in the backing cavity to form multiple-layer absorbers and thus extend the width of the absorption frequency band. The backing cavity can be partitioned into parallel cells using a honeycomb structure or it can be transformed to an irregular shape to increase the coupling between the acoustic modes in the backing cavity, thereby extending the sound absorption bandwidth. Even optimizing the coupling effect between the cavity acoustical response and the vibration of the MPP has been taken into account to expand the frequency bandwidth. Additional mechanical components have been proposed to be added to the simple panel-hole-cavity system to improve its performance using a flexible panel to substitute the rigid backing wall of a normal MPP system to produce a finite-sized micro-perforated panel-cavity-panel partition.

Despite all these efforts, the frequency range of MPP absorbers is still not satisfactory in many practical applications, especially for low frequency sound absorption applications where a larger cavity has to be used. Active control techniques have been introduced recently into MPP systems to address the low frequency limitations of the passive systems. In these cases, a loudspeaker and an error microphone have been placed behind an MPP layer to obtain the desired acoustic performance using "pressure releasing" on the back of the MPP or "impedance matching" of the active surface of the hybrid noise control system [22–25]. A hybrid passive-active system using flexible MPPs has been studied, where it has been shown that the panel vibration can increase the absorption at the structural resonance frequencies when the impedance matching approach is used [26].

The acoustic impedance of the hole of the micro-perforated panel can be approximated as [9]

$$Z_{\mathrm{MPP}} \approx \frac{32\eta t}{d^2}\left(\sqrt{1+\frac{K^2}{32}}+\frac{\sqrt{2}Kd}{32t}\right) - \mathrm{j}\rho_0\omega t\left[1+\left(1+\frac{K^2}{2}\right)^{-\frac{1}{2}}+0.85\frac{d}{t}\right] \tag{1.3.6}$$

where t is panel thickness, d is diameter of the hole, η is the dynamic viscosity coefficient, $\omega = 2\pi f$ is angular frequency and f is the frequency of interest, ρ_0 is the air density, and $K^2 = d^2\rho_0\omega/4\eta$. After obtaining the effective specific normal acoustic impedance Z_{MPPS} at the front surface of the MPP, and assuming that the MPP is local reactive, the complex amplitude reflection coefficient for a plane wave with an incidence angle of θ can be expressed as [19]

$$R(\theta) = \frac{Z_{\text{MPPS}} \cos \theta - \rho_0 c_0}{Z_{\text{MPPS}} \cos \theta + \rho_0 c_0} \tag{1.3.7}$$

Then the statistical absorption coefficient for random incidence can be obtained using

$$\alpha_{\text{st}} = 1 - 2 \int_0^{\pi/2} |R(\theta)|^2 \cos \theta \sin \theta \mathrm{d}\theta \tag{1.3.8}$$

1.4 Factors Affecting Sound Absorption

As introduced in the last two sections, the main mechanisms of sound absorbers are acoustic impedance matching on the absorbers' surface and acoustic energy dissipation inside the absorbers. If the acoustics impedance at the surface of the sound absorbers does not match that of the medium where the incident sound comes from, some or most of the sound is reflected back to the medium. When the acoustics impedance at the surface of the sound absorbers matches that of the medium, the sound is not reflected back to the medium, and this makes it possible to dissipate acoustic energy in the absorbers. To maximize the acoustic energy dissipation inside the absorbers, various physical effects can be used, such as viscous along the boundaries, heat exchanges, mechanical vibration, magnetic and electrical damping. For the textile materials, although a lot of factors such as fibre type, material thickness, fibre size, porosity, density, tortuosity and compression can affect the sound absorption of the material, there is no direct simple relationship between the value of the sound absorption coefficient and these factors [27].

Because the viscous and thermal effects are affected by the pore diameter, network shape and layout, density and other physical properties of the material in very complicated ways, the micromodels for understanding the underneath physical mechanisms of sound absorption in porous materials are quite complicated. Although there is no direct relationship between the physical and geometrical properties of textiles and the sound absorption, these micro effects for sound absorption of a uniform textile layer can be approximately represented by a macro quantity, i.e. the specific flow resistance. As introduced in Sect. 1.1.4, the flow resistivity of a textile layer is an intrinsic property of the textile and is a measure of how easily air can enter a porous textile material and the resistance that the air flow meets through the material. Equation (1.1.11) shows that the flow resistivity of certain fibreglass and rockwool is proportional to the dynamic gas viscosity linearly and the ratio of the porous material bulk density to the fibre material density nonlinearly, but is inversely proportional to the squared fibre diameter. The relationship between the flow resistance and sound absorption are very complicated. As shown by Eqs. (1.2.1)–(1.2.2) in Sect. 1.2, the characteristic impedance and propagation constant can be calculated with the flow resistivity in the nonlinear complex domain, which can then be used to obtain the specific acoustic impedance of the layer surface with Eq. (1.2.4) using complicated triangular functions. Finally, the absorption coefficient can be calculated from the ratio of the specific acoustic impedance of the layer surface to the characteristic impedance of the medium. In practice, the absorption performance of absorbers also depends on the sound field where it is located.

1.4.1 Material Thickness

Thickness of textile structures is one of the important parameters affecting the sound absorption. If the acoustics impedance at the surface of the textile structures matches that of the medium, the sound is not reflected back to the medium, then the thicker the structure is, the larger its sound absorption will be. To have effective sound absorption in the structure under this condition, the thickness of the structure should be at least one-tenth of the wavelength of the incident sound wave. This implies that thicker structures are required for absorbing low frequency sound due to its long wavelength. If the acoustics impedance at the surface of the textile structures partially matches that of the medium, thicker structures would absorb more sound that is not reflected back to the medium, especially in the low frequency range. If the acoustics impedance at the surface of the textile structures differs significantly with that of the medium, most sound is reflected back to the medium. Under this situation, the thickness of structures hardly affects sound absorption.

1.4.2 Fibre Size

As shown in Eq. (1.1.11), the flow resistivity is inversely proportional to the squared fibre diameter. This implies that finer fibres would result in high flow resistivity if all other parameters of the structure remain the same. Unfortunately, the relationship between the flow resistance and sound absorption is not straightforward but in a very complicated way. As shown by Eqs. (1.2.1)–(1.2.2), if the flow resistivity of the structures is very small, the characteristic impedance and propagation constant of the structure tend to be those in the medium. Under this situation, if there is no rigid wall against the structure and the structure is infinite thick and can dissipate the sound energy somehow, the sound absorption of the structure can be large. However, large fibre size does not necessarily result in small flow resistivity of the structures, and small flow resistivity of the structures does not necessarily result in large sound absorption. On the other hand, if the flow resistivity of the structures is very large, the characteristic impedance and propagation constant of the structure can be very different to those in the medium, so the sound absorption of the structure might be small.

1.4.3 Porosity

The porosity of a material indicates the amount of empty space or void present in the structure. Porosity is expressed as the ratio of amount of void present in the structure to the total volume of the sample

$$P_{\text{porosity}} = \frac{V_{\text{e}}}{V_{\text{t}}} \tag{1.4.1}$$

where V_{e} is the volume of the empty space and V_{t} is the total volume of the material. In the case of porous sound absorbers, the type, size and number of pores influence the sound absorption. Higher number of pores in a structure or large porosity usually means small porous material bulk density. As shown in Eq. (1.1.11), the flow resistivity is proportional to the porous material bulk density nonlinearly, which implies that large porosity would result in small flow resistivity if all other parameters of the structure remain the same. Unfortunately, the relationship between the flow resistance and sound absorption is very complicated. As shown in Eqs. (1.2.1)–(1.2.2), if the flow resistivity of the structures is small, the characteristic impedance and propagation constant of the structure tend to be those in the medium. Under this situation, if there is no rigid wall against the structure and the structure is infinite thick and can dissipate the sound energy, the sound absorption of the structure can be large. However, large porosity does not necessarily result in small flow resistivity of the structures, and small flow resistivity of the structures does not necessarily result in large sound absorption. On the other hand, if the porosity of the structure is small, the flow resistivity of the structures might be large, the characteristic impedance and propagation constant of the structure might differ from those in the medium significantly, so the sound absorption of the structure might be small. The methods of porosity measurement have been discussed in Chapter "Acoustical Test Methods for Nonwoven Fabrics".

1.4.4 Density

The density of a material indicates the mass concentration of the material, which is measured as mass per unit volume. In the case of porous structure, the bulk density plays an important role in acoustic absorption. As shown in Eq. (1.1.11), the porous material bulk density affects the flow resistivity nonlinearly. Usually, large density results in large flow resistivity if all other parameters of the structure remain the same. Unfortunately, the relationship between the flow resistance and sound absorption is very complicated. As shown in Eqs. (1.2.1)–(1.2.2), when the structure density is large, the flow resistivity of the structures is large, the characteristic impedance and propagation constant of the structure might differ from those in the medium significantly, and this results in small sound absorption of the structure; however, small density does not necessarily result in small flow resistivity of the structures, and small flow resistivity of the structures does not necessarily result in large sound absorption. For example, it has been found experimentally that the normal incidence sound absorption coefficient of a bamboo fibre material increases with its density [28].

1.4.5 Tortuosity

Tortuosity refers to the sinuosity and interconnectivity of void spaces in a porous structure. It describes the difference between the actual distance travelled by a fluid in a porous structure and the macroscopic travel distance. Tortuosity has several definitions depending on the field of study. Generally, tortuosity indicates the diffusion in porous media such as fibrous structures. Geometric tortuosity expressed as the ratio of the shortest path of interconnected points in pore fluid space to the straight distance between these points. In the case of acoustic materials tortuosity describes the elongation of the pathway through the pores. The methods used to measure tortuosity are described in Chapter "Acoustical Test Methods for Nonwoven Fabrics".

1.4.6 Compression

Textile structures used for acoustic absorption are compressible, which can alter the sound absorption property. When the textile structure or a porous material is compressed, the porosity and thickness values decrease, but the density increases. As shown in Eqs. (1.2.1)–(1.2.2), large structure density results in large flow resistivity of the structures if the other parameters of the structure remain the same. With a large flow resistivity, the characteristic impedance and propagation constant of the structure might differ from those in the medium significantly, and this results in low sound absorption of the structure; however, compression does not necessarily result in small sound absorption. Usually, compression increases sound absorption in the low frequency range at the cost of reducing sound absorption in the high frequency range. The methods used to measure compression are described in Chapter "Acoustical Test Methods for Nonwoven Fabrics".

1.4.7 Airflow Resistance

The acoustic absorption of a porous material is greatly influenced by the airflow resistance. The interlacement of threads in a woven structure or intermeshing of fibres in a nonwoven structure provides a resistance to the passage of air, which is an intrinsic property of the textiles. The airflow resistance is measured in terms of the resistance provided by the material per unit thickness. The details of flow resistivity have been described in Sect. 1.1.4.

1.4.8 Surface Impedance

The surface impendence of a structure is the short name of the specific boundary impedance of the structure, which is defined as the specific acoustic impedance that the boundary surface presents to an adjacent sound field, and its value is equal to the complex ratio of the acoustic surface pressure to the normal velocity of the fluid at the surface with the positive direction into the boundary [2]. As shown in Eq. (1.2.3), the absorption coefficient of a structure can be calculated from the ratio of the specific acoustic impedance of the layer surface to the characteristic impedance of the medium.

1.4.9 Position of Sound Absorbers

The effectiveness of sound absorbing materials depends on their relative placement. For acoustic materials used in buildings with various sizes and shapes, it is imperative to select the best locations for positioning of the acoustic material. Usually, the acoustic absorbers are placed along the edges or surfaces of the room for architecture or cost reasons; however, these positions might not be best locations for the porous material absorbers because the particle velocity of sound on the surfaces is usually very small. The details have been described in Sect. 1.1.2.

1.5 Applications of Sound Absorbers

Sound absorbers are commonly used in rooms to control sound field properties such as sound pressure level and reverberation time. To determine whether it is necessary to treat surfaces in a room with acoustically absorbing materials, the first step is to determine whether the reverberant sound field dominates the direct sound field at the point where it is desired to reduce the overall sound pressure level because treating reflecting surfaces with acoustically absorbing material can only affect the reverberant sound field [5]. At locations close to the sound source (for example, a machine operator's position) it is likely that the direct field of the sound source dominates, so it might be little improvement in treating a factory with sound absorbing material to protect operators from noise levels produced by their own machines. However, if an operator is affected by noise produced by other machines some distance away then treatment may be appropriate.

At any location in a room, the sound field is a combination of the direct field radiated by the source and the reverberant field. The acoustic energy due to the direct field at a location with a distance of r and in a direction θ from the source can be expressed by [3],

$$p_{\mathrm{d}}^2 = \frac{W\rho_0 c_0 D_\theta}{4\pi r^2} \tag{1.5.1}$$

where ρ_0 is the air density, c_0 is the speed of sound, W is the sound source power, D_θ is the directivity factor of the sound source. The acoustic energy due to the reverberant field can be expressed by [3],

$$p_{\mathrm{r}}^2 = \frac{4\rho_0 c_0 W(1-\alpha)}{S\alpha} \tag{1.5.2}$$

where S is the room surface area, and α is the total mean absorption coefficient (including air absorption) of the room. Thus the total acoustic energy measured at the location can be obtained by summation of the direct acoustic energy and the reverberant acoustic energy as

$$p^2 = \frac{W\rho_0 c_0 D_\theta}{4\pi r^2} + \frac{4\rho_0 c_0 W(1-\alpha)}{S\alpha} \tag{1.5.3}$$

It is obvious that the sound absorption only affects the reverberant acoustic energy. Written in the form of sound pressure level in dB, the above equation becomes

$$L_{\mathrm{p}} = L_{\mathrm{w}} + 10\log_{10}\left[\frac{D_\theta}{4\pi r^2} + \frac{4(1-\alpha)}{S\alpha}\right] + 10\log_{10}\frac{\rho_0 c_0}{400} \tag{1.5.4}$$

where L_{w} is the sound power level of the sound source.

The above equations can be used for designing sound pressure level control system in a room when the reverberant sound field dominates the direct field. Under this condition, the sound pressure level can be reduced or increased by adjusting the absorption added to the room. Assume the original total mean absorption coefficient of the room is α_0, but is changed to α_1 after control. The sound pressure level change due to the reverberant field can then be expressed as

$$\Delta L_{\mathrm{p}} = 10\log_{10}\frac{\alpha_1(1-\alpha_0)}{\alpha_0(1-\alpha_1)} \tag{1.5.5}$$

The reverberation time in a room can also be changed by adjusting its absorption as shown in the Sabine equation below [3],

$$T_{60} = \frac{55.25\,V}{c_0 S\alpha} \tag{1.5.6}$$

where S is the room surface area, V is the volume of the room, c_0 is the speed of sound and α is the total mean absorption coefficient of the room. The reverberation time T_{60} in a room can be increased by reducing the absorption in the room so the sound does not feel too dry or the reverberation time T_{60} in a room can be reduced by adding more sound absorbers in the room so the speech can be more clearly heard.

For example, a room of 8 m × 6 m × 3 m has a reverberation time of 0.64 s, and the sound source is located in the center of the room (not on ground) with a sound power level of 70 dB. It will be illustrated below, whether it is possible to reduce the reverberation sound level to 45 dB by putting some sound absorption material on the walls. The first step is to calculate the volume of the room and the room surface area, which are $V = 144\ \text{m}^3$ and $S = 180\ \text{m}^2$, respectively. Then the total mean absorption coefficient of the room can be calculated with the Sabine equation (1.5.6) as

$$\alpha = \frac{55.25V}{c_0 S}\frac{1}{T_{60}} = 0.2 \tag{1.5.7}$$

The sound pressure level of the direct sound 1 m away from the source, the reverberation sound, and the total sound 1 m away from the source can be calculated with Eq. (1.5.4).

$$L_{\text{pd}} = L_{\text{w}} + 10\log_{10}\frac{D_\theta}{4\pi r^2} + 10\log_{10}\frac{\rho_0 c_0}{400} = 59.2\ \text{dB} \tag{1.5.8}$$

$$L_{\text{pr}} = L_{\text{w}} + 10\log_{10}\frac{4(1-\alpha)}{S\alpha} + 10\log_{10}\frac{\rho_0 c_0}{400} = 59.6\ \text{dB} \tag{1.5.9}$$

$$L_{\text{p}} = 70 + 10\log_{10}\left[\frac{1}{4\pi} + \frac{4(1-0.2)}{180\times 0.2}\right] + 10\log_{10}\frac{415}{400} = 62.4\ \text{dB} \tag{1.5.10}$$

The reverberation sound level is 59.6 dB now and 14.6 dB is needed to be reduced to reach 45 dB. Substituting this number to Eq. (1.5.5), it has

$$10\log_{10}\frac{\alpha_1(1-\alpha_0)}{\alpha_0(1-\alpha_1)} = 14.6 \tag{1.5.11}$$

Substitute $\alpha_0 = 0.2$ to the equation and solve the equation, $\alpha_1 = 0.88$ can be obtained. Therefore, it is feasible to reduce the reverberation sound level to 45 dB by laying sound absorption material on the walls until the total mean absorption coefficient of the room reaches 0.88. However, this sound pressure level reduction is only significant at the locations where the reverberant sound dominates. As can be observed from Eqs. (1.5.8) and (1.5.10), the maximal total sound pressure level reduction at 1 m away from the source is approximately 3 dB because of the dominated direct sound.

1.6 Specific Discussions on Acoustic Textiles

A textile is a flexible woven material consisting of a network of natural or artificial fibres often referred to as thread, which is produced by spinning raw fibres of wool, flax, cotton or other material to produce long strands (https://en.wikipedia.org/wiki/Textile). Textiles are formed by weaving, knitting, crocheting, knotting or felting. The acoustic properties of textiles can be characterized in three categories as propagation, absorption, and scattering and those properties can be represented by flow resistance, transmission loss, absorption coefficient and scattering coefficient. The propagation and absorption properties of a textile depend on its flow resistance. Acoustic textiles are kinds of porous absorbers or resonant absorbers, so their acoustic properties can be described or predicted with what introduced in sections from 1.1 to 1.3.

Although theories for both membrane absorbers and micro-perforated plate absorbers have been well established, there is little method developed for their combination such as a fibreglass textile. Kang and Fuchs presented a theoretical method for predicting the absorption of such a structure [29]. The idea was to treat an open weave textile or a micro-perforated membrane as a parallel connection of the membrane and apertures. The predictions for both normal and random incidence showed reasonable agreement with measurements. It is found that the absorption coefficient of a fibreglass textile mounted at 100 mm from a rigid wall can exceed 0.4 over 3–4 octaves, and this range can extend to 4–5 octaves with two layers of the material over the same total air space of 100 mm.

The theoretical investigation into the sound propagation through flexible porous media is of prime importance for evaluating the noise absorption capacities of foam materials or textiles, such as woven fabrics or nonwoven fibrewebs [30]. The Zwikker and Kosten model [31] for sound propagation through porous flexible media has been used for numerical calculations of some intrinsic characteristics of nonwoven fibrewebs so that it can yield the highest sound absorption coefficients in the audible frequency range. These results can serve as guidelines for the optimal design of acoustic elements made of textile assemblies.

Micro-fibre fabric has fine fibres and a high surface area and it has been used in such applications as wipers, thermal insulator, filters or breathable layers, as well as for sound absorption. The feasibility of using micro-fibre fabrics as sound absorbent materials has been investigated [32]. The test results of five micro-fibre fabrics and one regular fibre fabric showed that the micro-fibre fabrics' sound absorption is superior to that of conventional fabric with the same thickness or weight, and the micro-fibre fabrics' structure was found to be important for controlling sound absorption according to sound frequency. Furthermore, fabric density was found to have more effect than fabric thickness or weight on sound absorption, and the Noise Reduction Coefficient increases to its highest value at a fabric density of about 0.14 g/cm^3, and it decreases thereafter.

A review paper was published to describe how the physical prosperities of materials like fibre type, fibre size, material thickness, density, airflow resistance and porosity can change the absorption behaviour [27]. The effect of surface impedance, placement of sound absorptive and compression, on sound absorption behaviour of materials was also considered. The results showed the relationship among the sound absorption and airflow resistance, material thickness, air gap and attachment film, and it was found that higher airflow resistance usually gives better sound absorption values under a certain value, the creation of air gap behind the absorptive material increases sound absorption coefficient values, as discussed in Sects. 1.2 and 1.3.

Nonwoven fabrics can be used in car interior components (head liners, doors, side panels and trunk liners) to prevent noise from reaching the passenger compartment and so achieving comfort in the car interior [33]. Two kinds of fibres (polyester and hollow polyester fibres, both six denier) were used to produce three different fabrics (100 % polyester fibres, 75 % polyester/25 % hollow polyester fibres and 55 % polyester/45 % hollow polyester fibres) with four fabric weights (300, 400, 500 and 600 g/m^2). It was found that the samples produced with high percentage of hollow fibres had high sound absorption, whereas samples produced with 100 % polyester fibres had the lowest rates. It was also found that there are direct relationship between weight per m^2 and sound absorption efficiency. The sample produced with 55 % polyester/45 % hollow polyester fibres and 600 g/m^2 has the best absorption.

The sound absorption properties for recycled fibrous materials have been investigated, which include natural fibres, synthetic fibres and agricultural lignocellulosic fibres [27]. The results indicated that nonwoven samples had high sound absorption coefficients at high frequencies (2000–6300 Hz), low sound absorption coefficients at low frequencies (100–400 Hz) and better sound absorption coefficients at mid (500–1600 Hz) frequencies, and could be improved by increasing the thickness of nonwovens. The rice straw and sawdust composite samples have low sound absorption at low and mid frequencies. The sound absorbing performance can be improved by adding perforation of 6 % for the tested sample, increasing the thickness of nonwoven samples and adding air spaces behind the tested composite systems.

1.7 Conclusions and Future Directions

As shown in this chapter, sound absorbers can be made from porous materials or resonant structures, and the main mechanisms are acoustic impedance matching on the absorbers' boundary and acoustic energy dissipation inside the absorbers. If the acoustics impedance at the boundary of the sound absorbers does not match that of the medium where the incident sound propagates, some or most of the sound will be reflected back to the medium. Under this situation, no matter how efficient the absorption mechanism inside the absorbers is, the reflected acoustic energy cannot be dissipated, so the sound absorption coefficient cannot be large. When the acoustics impedance at the boundary of the sound absorbers matches that of the medium, the sound will not be reflected back, and this makes it possible for dissipating acoustic energy in the absorbers. To maximize the acoustic energy dissipation inside the absorbers, various physical effects can be used, such as viscous along the boundaries, heat exchanges, mechanical vibration, magnetic and electrical damping.

Even with all the progress and so many kinds of sound absorbers, there are still many aspects that can be improved for better sound absorption. One of the challenges is to reduce the size of the absorbers for low frequency absorption. For example, a micro-perforated panel needs to be backed by a 0.92 m deep cavity to have good absorption for 100 Hz sound because the wavelength of the 100 Hz is about 3.4 m in air. An integrated mechanical and electrical sound absorber based on Micro-Perforated Panels and Shunted Loudspeakers (MPPSL) has been proposed to reduce the size of the absorber [34]. It is expected a 0.1 m thick MPPSL should be able to have better sound absorption performance than the micro-perforated panel backed by a 0.92 m deep cavity for 100 Hz sound.

The idea of the MPPSL is to integrate mechanical and electrical components into acoustics absorbers to reduce the size of devices for low frequency sound control. Because the wavelength of low frequency sound is quite long, the size of normal Micro-Perforated Panels cannot be reduced significantly if only the acoustical energy dissipation mechanism is engaged. Mechanical energy dissipation components such as thin panels are useful; however, they are not as versatile and powerful as electrical energy dissipation components. It is the powerful energy storing and dissipating capability and design flexibility of the electrical components that make it possible to use compact devices to control long wavelength sound.

Another challenge for future sound absorbers is smart absorbers that can change its absorption coefficient under different situations. Active absorbers might be a solution to the problem, which uses microphones and loudspeakers to interact with the original sound field with active control technologies [4]. Active noise control is a method for reducing existing noise via the introduction of controllable secondary sources to affect generation, radiation, transmission and reception of the original primary noise source. It can provide better solutions to low frequency noise problems than current passive noise control methods when there are weight and volume constraints. As an active sound absorber, the system can adjust the sound absorption coefficient according to the need, and it offers the possibility of bass absorption or diffuse reflection from relatively shallow surface as well as the possibility of variable acoustics.

References

1. Qiu X, Lu J, Pan J (2014) A new era for applications of active noise control. In: Proceedings of the 43rd international congress & exhibition on noise control engineering. Melbourne, Australia
2. Morfey CL (2001) Dictionary of acoustics. Academic Press

3. Bies DA, Hansen CH (2009) Engineering noise control—theory and practice. 4th edn. Spon Press
4. Cox TJ, D'Antonio P (2009) Acoustic absorbers and diffusers: theory, design and application. 2nd edn. Taylor and Francis
5. Kuttruff H (2009) Room acoustics. 5th edn. Taylor & Francis
6. Delany ME, Bazley EN (1970) Acoustical properties of fibrous absorbent materials. Appl Acoust 3(3):105–116
7. Kinsler LE, Frey AR, Coppens AB, Sanders JV (2000) Fundamentals of acoustics. 4th edn. John Wiley and Sons Inc.
8. Pierce AD (1981) Acoustics. McGraw-Hill Book Company
9. Maa DY (1998) Potential of microperforated panel absorber. J Acoust Soc Am 104:2861–2866
10. Fuchs HV, Zha X (2006) Micro-perforated structures as sound absorbers—a review and outlook. Acta Acust Acust 92:139–146
11. Toyoda M, Takahashi D (2008) Sound transmission through a microperforated-panel structure with subdivided air cavities. J Acoust Soc Am 124:3594–3603
12. Bravo T, Maury C, Pinhède C (2012) Sound absorption and transmission through flexible micro-perforated panels backed by an air layer and a thin plate. J Acoust Soc Am 131 (5):3853–3863
13. Bravo T, Maury C, Pinhède C (2012) Vibroacoustic properties of thin micro-perforated panel absorbers. J Acoust Soc Am 132(2):789–798
14. Liu J, Herrin DW (2010) Enhancing micro-perforated panel attenuation by partitioning the adjoining cavity. Appl Acoust 71:120–127
15. Ruiz H, Cobo P, Jacobsen F (2011) Optimization of multiple-layer microperforated panels by simulated annealing. Appl Acoust 72:772–776
16. Ruiz H, Cobo P, Dupont T, Martin B, Leclaire P (2012) Acoustic properties of plates with unevenly distributed macroperforations backed by woven meshes. J Acoust Soc Am 132 (5):3138–3147
17. Toyoda M, Mu RL, Takahashi D (2010) Relationship between Helmholtz resonance absorption and panel-type absorption in finite flexible microperforated panel absorbers. Appl Acoust 71:315–320
18. Wang CQ, Huang L (2011) On the acoustic properties of parallel arrangement of multiple micro-perforated panel absorbers with different cavity depths. J Acoust Soc Am 130(1):208–218
19. Wang CQ, Cheng L, Pan J, Yu GH (2010) Sound absorption of a micro-perforated panel backed by an irregular-shaped cavity. J Acoust Soc Am 127:238–246
20. Yang C, Cheng L, Pan J (2013) Absorption of oblique incidence sound by a finite micro-perforated panel absorber. J Acoust Soc Am 133(1):201–209
21. Zou J, Shen J, Yang J, Qiu X (2006) A note on the prediction method of reverberation absorption coefficient of double layer micro-perforated membrane. Appl Acoust 67:106–111
22. Cobo P, Cuesta M (2007) Hybrid passive-active absorption of a microperforated panel in free field conditions. J Acoust Soc Am 121:EL251–EL255
23. Cobo P, Pfretzschner J, Cuesta M, Anthony DK (2004) Hybrid passive-active absorption using microperforated panels. J Acoust Soc Am 116:2118–2125
24. Leroy P, Berry A, Herzog Ph, Atalla N (2011) Experimental study of a smart foam sound absorber. J Acoust Soc Am 129(1):154–164
25. Zou H, Qiu X, Lu J, Li N (2013) A study of a hybrid pressure-release sound absorbing structure with feedback active noise control system. In: Proceedings of the 42nd international congress & exhibition on noise control engineering. Innsbruck, Austria
26. Zheng W, Huang Q, Li S, Guo Z (2011) Sound absorption of hybrid passive-active system using finite flexible micro-perforated panels. J Low Freq Noise Vib Act Control 30(4):313–328
27. Seddeq HS (2009) Factors influencing acoustic performance of sound absorptive materials. Aust J Basic Appl Sci 3(4):4610–4617

28. Koizumi T, Tsujiuchi N, Adachi A (2002) The development of sound absorbing materials using natural bamboo fibers, high performance. WIT Trans Built Environ 59:1–10
29. Kang J, Fuchs HV (1999) Predicting the absorption of open weave textiles and micro-perforated membranes backed by an air space. J Sound Vib 220(5):905–920
30. Shoshani Y, Yakubov Y (2000) Numerical assessment of maximal absorption coefficients for nonwoven fiberwebs. Appl Acoust 59(1):77–87
31. Zwikker C, Kosten CW (1949) Sound absorbing materials. Elsevier Publishing Company, Oxford
32. Na Y, Lancaster J, Casali J, Cho G (2007) Sound absorption coefficients of micro-fiber fabrics by reverberation room method. Text Res J 77(5):330–335
33. Mahmoud AA, Ibrahim GE, Mahmoud ER (2011) Using nonwoven hollow fibers to improve cars interior acoustic properties. Life Sci J 8(1):344–351
34. Tao J, Jing R, Qiu X (2014) Sound absorption of a finite micro-perforated panel backed by a shunted loudspeaker. J Acoust Soc Am 135:231–238

Materials Used for Acoustic Textiles

Asis Patnaik

Abstract This chapter discusses about textile materials used for noise/sound absorption applications. Various types of raw materials used for manufacturing of acoustic materials such as, natural and synthetic fibres, along with recycled materials were covered. Different types of porous absorbing materials as well as panel absorbing materials were discussed. Acoustic properties of materials coated, sprayed or embedded with chemicals and fillers were also covered. Use of nanomaterials for acoustic applications particularly for absorbing low and medium frequencies were discussed. Some of the recent trends in developing hybrid type of acoustic absorbers where acoustic property was effectively controlled with the help of active as well as passive absorbers were also covered. Furthermore, new types of panel absorbers were also discussed.

Keywords Acoustic materials · Nonwovens · Porous absorber · Panel absorber · Sound absorption

1 Introduction

Noise/sound absorption plays an important role in human well-being as it has a direct impact on health and productivity levels. Unwanted noise can be reduced or eliminated by the use of sound absorbing materials, barriers and enclosures of the

A. Patnaik (✉)
CSIR Materials Science and Manufacturing, Polymers and Composites Competence Area, P.O. Box 1124, 6000 Port Elizabeth, South Africa
e-mail: apatnaik@csir.co.za; asispatnaik@gmail.com

A. Patnaik
Department of Textile Science, Faculty of Science, Nelson Mandela Metropolitan University, P.O. Box 77000, 6031 Port Elizabeth, South Africa

A. Patnaik
Faculty of Engineering, Department of Clothing Management and Textile Technology, Cape Peninsula University of Technology, P.O. Box 1906, Bellville campus, 7535 Cape Town, South Africa

R. Padhye and R. Nayak (eds.), *Acoustic Textiles*, Textile Science and Clothing Technology, DOI 10.1007/978-981-10-1476-5_4

sound source. Sound absorbing materials are used to absorb most of the sound energy when sound waves encounter with them. Depending upon the ability of the sound absorbing material, some part of the energy of the sound waves is dampened through the frictional contact with the fibres and remaining part of the energy is transmitted through the material. Materials which can absorb most of the incident sound energy and effectively control the noise are known as sound absorbing or acoustic materials. The sound absorbing materials can be used at different locations such as close to the sound source, in the path of the sound and close to the receiver to control the noise.

While designing the acoustic material for various applications, the primary aim is to achieve desired sound absorption properties in the required sound frequency range. The same material may behave differently with different sound frequencies. Hence for practical applications, the sound absorbers should be selected on the basis of the sound frequency. Furthermore, there may be secondary requirements such as light reflectivity, flame retardancy, durability, etc., which sound absorbing material need to satisfy.

A wide range of materials such as porous, foam, panel can be used for sound absorption applications. It has been established that the structures with high porosity are ideal candidates for effective sound absorption. A porous material consists of void spaces which help the sound wave to enter into the structure. Textile structures or fibrous materials are a good example of porous material used for sound absorption. There are number of textile materials used for acoustic applications. They can be classified as woven, nonwoven and knitted sound absorbers. Nonwovens are the most widely used for acoustic applications. Designing nonwovens with good aesthetic appeal is a challenging job. Hence sometimes they are used in combination with woven or knitted fabrics, depending upon required applications.

Both natural and synthetic fibres, along with recycled materials are used to manufacture textile-based acoustic materials. These materials are either in the form of a porous absorber or a panel absorber for acoustic applications. At times, textile-based acoustic materials are coated and sprayed with speciality chemicals in order to improve various performance properties as per the required applications.

Noise/sound absorption coefficient (NAC) and sound transmission loss (STL) are the two terminologies used to define acoustic properties of a material. If the material is porous, then NAC is used to define its acoustic property. Most of the textile materials' acoustic properties are defined in terms of NAC values. For example, automotive carpet, door liners, dash insulators, form such class of materials that are in the form of nonwoven mats. If the material has a flat surface, like panel, metal plate, then STL is used to define its acoustic property.

This chapter discusses about textile materials used for sound absorption applications including various types of raw materials used to produce such acoustic materials. Different types of porous and panel absorbing materials as well as chemically coated materials were discussed. Application of nanomaterials and hybrid type of acoustic absorbers were also covered.

2 Raw Materials

2.1 Natural and Synthetic Fibres

Natural and synthetic fibres along with recycled materials like shoddy are widely used for the manufacturing of acoustic materials. Noise control can be achieved in the form of an acoustic absorber or barrier material by blending natural and synthetic fibres, either by using synthetic fibres alone or through blending post-consumer or industrial waste like shoddy with natural and synthetic fibres (Fig. 1). Most of the textile materials are widely used in the form of acoustic absorber.

The application of natural fibres in acoustic control is now gaining impetus due to the environmental concerns associated with synthetic materials [2, 3]. Some of the natural fibres used for acoustic applications are, flax, hemp, kenaf, jute, agave, bamboo, coir (plant based), and wool (animal based) [2, 4, 3]. Examples of the synthetic fibres used in acoustic applications are polypropylene (PP), polyester (PET), viscose, bi-component, etc.

The fibre properties controlling the acoustic absorption in various applications include fibre length, fineness, tenacity and uniformity. Selection of right type of material depending on the field of use and durability is the key challenge while designing acoustic materials. One of the important fibre parameters affecting the acoustic properties is the fibre diameter. The diameter of natural fibres is generally greater than the synthetics fibres. Furthermore, there may be large variations in fibre diameter and cross section in natural fibres as compared to the synthetic fibres. If the acoustic structure contains finer and regular diameter, it is easier to control the acoustic properties. On the other hand, it is rather difficult to design acoustic structures from coarser fibres with variable diameter.

One of the requirements for acoustic application are acceptable strength and low elongation properties. Hence, fibres such as flax, hemp, jute and coir are the leading

Fig. 1 Photograph of various types of acoustic materials (*Source* CSIR [1])

fibres for acoustic applications. Furthermore, these fibres are available in abundance in different countries around the world. Similarly fibres like PP, PET, are widely used for acoustic applications because of the fibre uniformity and good mechanical properties.

There are certain advantages and limitations while selecting natural and synthetic materials or their blends for acoustic applications. The advantages and disadvantages of using natural fibres over their synthetic counterpart are discussed in the following section. The advantages of using natural fibres in acoustic applications include:

- Low level of pollution,
- Reduced carbon emission,
- Safer than glass fibres as these fibres are free from respiratory diseases or skin irritation, and
- Reduced waste as natural fibres are biodegradable.

The disadvantages of using natural fibres in acoustic applications include,

- Lower tenacity,
- High variability in the fibre properties depending on the geographical location and weather,
- High moisture regain properties cause swelling of the fibre,
- Limitations in terms of durability and flame retardancy, and
- High variation in the prices of the raw materials.

In order to overcome the above disadvantages of natural fibres, they are blended with synthetic fibres. Although synthetic fibres have several advantages over natural fibres, their environmental concern is very high as they are produced by high temperature extrusion from chemicals, often based on petrochemical sources. Furthermore, the natural fibres are safer than synthetic fibres during handling and disposal stages. Other performance properties of natural materials can be improved by various processes such as impregnation, coating, carbonization and drying to make them equivalent (to some extent) to the synthetic materials. For example, fire and moisture resistance properties are required for natural fibre-based materials used for acoustic applications in automotive, building and aerospace industries. The fire and moisture resistance properties can be improved by treating the fibres with flame retardants and moisture resistant finishes, respectively.

In addition to the above advantages and disadvantages related to natural and synthetic fibres, the mineral fibres (rockwool and fibreglass) used for acoustic applications are associated with health hazards. These fibres are carcinogenic to human and can cause respiratory problems. Hence, these fibres are gradually being replaced with the synthetic or blends of synthetic and natural fibres.

Thilagavathi et al. [5] studied the acoustic properties of natural fibre nonwovens made from bamboo, banana and jute fibres blended with PP fibres in 50:50 blends for automotive interior applications. The authors have studied the NAC values in the frequency range of 100–1600 Hz. NAC values of bamboo/PP was the highest as compared to banana/PP and jute/PP in the above frequency range. At 1250 Hz, the NAC values for bamboo/PP, banana/PP and jute/PP were 0.20, 0.13 and 0.17, respectively.

Parikh et al. [6] studied the acoustic properties of needle-punched natural fibre nonwovens for automotive floor covering systems. Natural fibres (kenaf, jute, waste cotton and flax) were blended with PET and PP in 35/35/30 ratios, whereas PET/PP were blended in 70/30 ratios. The area weights of the samples were in the range of 730–1062 g/m^2, whereas, the thickness were in the range of 8–10.10 mm. Authors evaluated NAC values in the frequency range of 100–3200 Hz and observed that with the addition of a cotton and poly urethane (PU) under pad along with the above nonwoven blends increased the NAC values. For a kenaf/PET/PP blend, the NAC values at 700, 1200, 2200 and 3200 Hz were 0.07, 0.14, 0.31 and 0.54, respectively. The corresponding NAC values at above frequencies with cotton under pad were 0.13, 0.27, 0.59 and 0.81, respectively. Similarly, the NAC values for PU under pad were 0.28, 0.54, 0.89 and 1.00, respectively. These results show that, with the addition of an under pad, the NAC values increases.

In another recent study, flax, hemp and agave fibres were blended with PET fibres in 50/50 proportions and acoustic properties of the needle-punched mats were evaluated [2]. A maximum NAC value of 0.45 was obtained from the selected parameters of fibres, depth of needle penetration and stroke frequency. These NAC values were obtained when above three factors were acting simultaneously during the mat production stage. The authors have also observed that when multiple factors were acting simultaneously during mat production stage, fibre type played the least dominant role in determining sound absorption properties.

Acoustic properties of materials made from cotton, flax, ramie, sisal, jute and wool fibres were reported at 500 Hz [7]. The NAC values at 500 Hz were 0.50, 0.40, 0.40, 0.10, 0.20 and 0.20, respectively, for cotton, flax, ramie, sisal, jute and wool fibre based absorbers.

Patnaik et al. [3] studied the sound insulation properties of needle-punched samples developed from waste wool and recycled polyester fibres (RPET) for building industry applications. Short wool fibres obtained by shearing the sheep hair are not suitable for apparel/clothing purpose; and therefore are generally discarded as a waste material. These wool fibres are also obtained from the sheep's nurtured for meat production. These non-standard fibres are known as waste wool fibres. Waste wool fibre is a potential source of raw material for thermal and sound insulation applications, but its quantities are limited. In order to overcome above problem, waste wool fibres were mixed with RPET fibres to prepare needle-punched mats. Two types of waste wool fibres, coring wool (CW) and dorper wool (DW) were used. These names were derived from the sheep breed. Two nonwoven mats were produced from these fibres. Another mat was produced from RPET fibres. Waste wool fibres were then combined with RPET in 50/50 proportions in the form of a two layer mat and named as coring wool product (CWP) and dorper wool product (DWP) as shown in Table 1. In total, five samples were produced and their acoustic properties were evaluated apart from other performances properties. The nominal area weight of the samples (mats) was

Table 1 Sample compositions and their physical properties (*Source* Patnaik et al. [3])

Sample code	Sample composition	Number of layers	Thickness (mm)	Bulk density (kg/m^3)
Coring wool (CW)	100 %	Single	15	66.66
Dorper wool (DW)	100 %	Single	17	58.82
Recycled polyester (RPET)	100 %	Single	16	62.50
Coring wool product (CWP)	50 % (CW)	Double	16	62.50
	50 % (RPET)			
Dorper wool product (DWP)	50 % (DW)	Double	17	58.82
	50 % (RPET)			

Note The number shows the % of fibres by weight

Fig. 2 Photograph of the samples: **a** DW 100 %, **b** CW 100 %, **c** RPET 100 %, **d** DWP, **e** CWP (*Source* Patnaik et al. [3])

1000 g/m^2. Thicknesses of the samples were in between 15 and 17 mm. Photographs of the samples are shown in Fig. 2.

All developed samples showed good sound absorption properties in the overall frequency range (50–5700) Hz. NAC values of the samples in various frequency ranges are shown in Table 2. The sound absorption was lower at low frequencies (50–1000 Hz), and increased from medium (1000–2000 Hz) to high frequency range (200–5700 Hz) for all the samples Patnaik et al. [3].

The lowest NAC value was 0.61 for RPET and the highest was 0.75 for DWP in the overall frequency range (50–5700 Hz). Sound absorption depends upon thickness of the material amongst other factors. Higher the NAC value, better will be the sound absorption property. The reason can be attributed to the fact that the kinetic energy of the incident sound wave gets converted to low level of heat energy when it passes through a thicker structure. Thicker structure absorbs sound wave by causing frictional loss between sound wave and fibres, thereby dampening the effects of the propagating sound wave. Another factor was the tortuosity of the nonwoven mats, which is a ratio of the length of the open pores (i.e. the length of

Table 2 Acoustic coefficients (α) in various frequency ranges (*Source* Patnaik et al. [3])

Sample code	50–1000 (Hz)	1000–2000 (Hz)	2000–5700 (Hz)	50–5700 (Hz)
CW	0.10 (0.050)	0.42 (0.050)	0.85 (0.057)	0.69 (0.058)
DW	0.16 (0.064)	0.55 (0.062)	0.94 (0.062)	0.74 (0.063)
RPET	0.09 (0.012)	0.34 (0.015)	0.81 (0.025)	0.61 (0.029)
CWP	0.13 (0.032)	0.48 (0.031)	0.89 (0.030)	0.71 (0.032)
DWP	0.18 (0.036)	0.58 (0.034)	0.95 (0.033)	0.75 (0.035)

Values in the parenthesis indicate the standard deviation

the interconnected through pores) to the material thickness. Higher thickness of the sample allows the sound wave to be channelled through the tortuous path of the through pores, thereby creating more frictional losses and thereby increasing the NAC value. In the case of lower thickness, path lengths available for the propagation of the sound wave were not enough to cause sufficient amount of frictional loss. Also, DWP was a two layer structure, designed to entrap the sound wave within the structure. Two layer structure with air gap between them assisted in dampening the sound wave within the sample [3].

Cheng and Jiang [8] reported the improvement in acoustic properties of layered nonwoven composites. Layered nonwoven composites were prepared by combining activated carbon surface layer with the needle-punched base layers of cotton, ramie and PP. One of the reasons for improvement in acoustic properties of layered nonwovens was due to the presence of small pore size created by the carbonization technique.

2.2 Recycled Materials: Shoddy

Post-consumer or industrial waste materials like shoddy are widely used in acoustic applications. The main reasons for using such materials are their cost advantages and acceptable acoustic and other performance properties. Some of the disadvantages of using the shoddy are variation in acoustic properties of the materials in some places, as some offcuts of yarns from fabrics are still there in the shoddy instead of fibres. These yarns lead to significant variation in material thickness as well as acoustic properties. Furthermore, the cost of energy spent in recycling shoddy materials are also increasing.

While using the shoddy, it is easier to process loose structure such as knits as compared to the tight woven structure and less energy is spent in the material conversion. Acoustic properties of recycled materials also depend on the thickness of the material. However, as mentioned earlier there may be wider variation of the acoustic properties for recycled absorbers.

The global consumption of textiles have increased manifold compared to what it was a decade ago [9]. The recent consumers prefer frequent style changes leading to a range of products to their end-of-life. Hence, million tonnes of textile waste is generated annually around the globe. It is essential to recycle these materials to reduce the environmental loading and achieve economic benefits. The recycled acoustic materials can be prepared from various end-of-life products or wastes such as apparel cutting waste, nonwoven waste, woven selvages and other textile waste.

Lee and Joo [10] studied the acoustic properties of thermal bonded nonwovens made up of RPET fibres. In one set of experiment, they compared NAC values of four different nonwovens, each having 40 % low melt PET fibres (6 denier, 42 mm) as one of the blend component. Other blend component consisted of regular PET fibres in following proportions: 60 % (2 denier, 38 mm), 20 % (1.25 denier, 38) and 40 % (2 denier, 38 mm), 40 % (1.25 denier, 38) and 20 % (2 denier, 38 mm) and 60 % (2 denier, 38 mm). As the fine fibre contents were increased from 0 to 60 %, NAC values were increased up to 1000 Hz, after that there were no changes in the NAC values. In another experiment, they varied the web orientation angle (0°, 35°, 45° and 90°) during the carding process. Higher NAC values (0.95–0.98) were observed from 1000 to 4000 Hz for all the samples, although there was not much difference between the samples. The web which was laid at 90° showed the highest NAC among the samples.

Manning and Panneton [11] studied the acoustic properties of shoddy-based absorbers (post-consumer or industrial recycled fibres). Shoddy mainly consisted of PET (32 %), cotton (28 %), acrylic (18 %), PP (6 %), wool (8 %), nylon (3 %) and other (3 %) fibres. They compared three different shoddy samples, needle-punched mat (area weight 986 g/m^2 and thickness 10.6 mm), thermally bonded mat (area weight 838 g/m^2 and thickness 12.7 mm) and resin bonded mat (area weight 729 g/m^2 and thickness 13.5 mm). There was not much difference in the NAC values of the samples in frequency range of 0–4000 Hz. A lower NAC value of 0.20 was obtained at low frequency (0–1000 Hz) for all samples.

Acoustic properties of various types of needle-punched mats including recycled and natural fibres have been reported by Seddeq et al. [12]. They studied PET/cotton/wool (70/30), PP/cotton/wool (80/20), cotton/PET (50/50) blends and 100 % jute fibres. The area weights of the samples were in the range of 422–561 g/m^2, whereas the thicknesses were in the range of 2.53–5.64 mm. The NAC values were low (0.06) at low frequencies which authors defined in that work as 100–400 Hz. The NAC values were higher (up to 0.67) at high frequencies (2000–6300 Hz). Since NAC values were lower at low frequencies, authors suggested that it can be improved by increasing the thickness of the samples.

3 Porous Absorbers

Porous absorbers are the most commonly used absorbing materials. A porous absorbing material or porous absorber is a solid that contains cavities, channels or pores so that sound waves are able to enter through them [13]. Porous absorbers can be in the form of fibrous, cellular or granular materials. Fibrous materials are textile structures prepared from natural, synthetic and natural–synthetic blends. Cellular materials are in the form of foam and granular materials are in the form of asphalts, clay, concrete and soil. The example of cellular materials includes foams and open-celled PU.

This section mainly focuses on fibrous materials used for acoustic absorption. Some of the fibrous porous absorbers were already covered in Sect. 1. Majority of the commercial porous absorbers are fibrous. These materials consist of fibres or a set of continuous or staple yarns with entrapped air in the structure. Fibrous absorbing materials are generally manufactured by weaving, nonwoven and knitting technologies. Among these, nonwoven materials dominate the manufacturing technology, particularly the needle-punched nonwoven materials. The following points describe the reason for wide application of nonwovens:

- The properties can be controlled by changing the relationship between fibre–fibre rather than yarn–yarn in the case of other two,
- Nonwovens are easy to produce and cost-effective as several processes related to spinning and weaving are eliminated for nonwovens,
- Wide range of fibres can be processed,
- Several nonwoven manufacturing technologies can be combined together.

Most of the widely used porous fibrous absorbers are carpet, under carpet insulator, ceiling insulator, dash insulator, drapery and door panel insulator. Carpet is an important material for sound absorption. The effectiveness of a carpet depends on its construction, backing material, pile features, thickness, underlay and weight. The pile features include its height, density, structure, fibre content and air gap [14, 15]. In addition, type of backing material (coated and uncoated), type of pile (loop or cut), type and thickness of the pad affect the sound absorption.

The sound absorption of a drapery depends on its airflow resistance and its distance from the window or wall [16]. The greater is the distance from the wall, the better is the sound absorption at low frequencies. Furthermore, the curtain configuration can also affect the sound absorption. The curtains in a folded form serve as better sound absorbers as compared to when they are flat.

The porous absorbers can be classified as sheet materials and bulk materials depending on their thickness with respect to the wavelength of sound [17]. When the thickness of the material is much smaller than the wavelength of sound they are called sheet materials. When the thickness is much larger than the wavelength, they are called bulk materials. The acoustic absorption of sheet materials depends on the viscous effects and areal density (mass per unit area), whereas it depends on the material density, viscous and thermal effect in case of bulk materials. The porosity of textile structures can be contributed by the interyarn/fibre porosity (as in nonwovens) or both interyarn/fibre and intrayarn/fibre porosity (woven and knits).

Kucuk and Korkmaz [18] studied various combinations of fibres and different nonwoven bonding techniques on the acoustic properties of mats and compared them with a commercial automotive mat. Thermal, spunlacing and needle-punching bonding techniques were used to prepare various types of mats using different fibre blends. Thermal bonding was used to prepare mats utilizing 70 % wool and 30 %

bi-component PET, 70 % cotton and 30 % PET, 70 % acrylic–cotton–PET and 30 % PP, 90 % PET and 10 % low melt PET fibres. The area weights were in the range of 550–2900 g/m^2. The thicknesses of the materials were in the range of 2.47–35.38 mm. The area weight and thickness of the commercial needle-punched reference mat was 712 g/m^2 and 16.38 mm, respectively. By comparing NAC values of various samples, as compared to commercial mat, thermal bonded 70 % cotton and 30 % PET showed the best result amongst all samples in mid to high frequency range of 1000–6300 Hz, with a maximum NAC value of 0.96 obtained at 3000 Hz. The corresponding value for commercial mat obtained at 3000 Hz was 0.35. The reason for such increase was higher area weight (almost double) and thickness of those mats that helped in absorbing more sound waves.

The spunlaced nonwovens produced from 70 % PET and 30 % polyamide and 100 % PET fibres, were having area weights in the range of 45–170 g/m^2 and thicknesses in the range of 0.23–0.83 mm. When NAC values of these nonwovens were compared with the commercial mat, then this mat was a superior material in the entire frequency range of 50–6300 Hz [18]. Similarly, on comparing the NAC values of the developed needle-punched nonwovens (100–900 g/m^2, 0.83–7.34 mm, 100 % PET and meta-aramid) with the commercial mat, the 900 g/m^2 showed slightly better acoustic property as compared to the commercial mat. Higher NAC values were observed at higher frequency for both mats. Whereas lower area weight needle-punched mats (100–500 g/m^2), showed lower NAC values in entire frequency range 50–6300 Hz as compared to the commercial mat.

Acoustic properties of several nonwovens made from different types of materials were studied by Yang and Yu [19]. The fibre used for this study were PET, PET/viscose blend (70/30), glass and basalt fibres. The area weights were in the range of 42–581 g/m^2. The thicknesses of the materials were in the range of 0.523–5.275 mm. The NAC values were increased with increasing area weight and thicknesses of the materials. Maximum NAC values were obtained for glass fibre based nonwoven in the frequency range of 800–6200 Hz.

Acoustic absorptive properties of spun-bonded nonwovens made from islands-in-the-sea bi-component filaments were reported by Suvari et al. [20]. The authors studied the sound absorption behaviour by passing the spun-bonded nonwovens thrice through a spun lacing machine. A three- and four-layered 108 islands-in-the-sea were compared with a trough air bonded high loft nonwoven for acoustic properties. The area weights of high loft nonwoven and three and four-layered 108 islands-in-the-seas were 350, 411 and 548 g/m^2, respectively. The thicknesses of high loft nonwoven and three- and four-layered 108 islands-in-the-seas were 5, 1.32 and 1.85 mm, respectively. Authors observed a higher NAC values for three and four layered 108 islands- in-the-seas as compared to high loft nonwoven in the frequency ranges, 2032–4600 Hz and 2808–5322, respectively.

Bo and Tianning [21] investigated the acoustic absorption characteristics of porous sintered fibre metal. The porous metal provides the benefits of both metal (ductability, conductivity, heat transfer and high specific stiffness) and porous materials (filtration, noise absorption and energy absorption). The metal porous structures are suitable for application in extreme conditions compared to conventional fibrous porous materials.

4 Panel Absorbers

This section discusses about panel absorbers in which several materials are used in combinations to achieve the desired acoustic insulation. In most cases, panel absorbers are materials having flat surface with holes in it. Gypsum, extruded polystyrene, magnesium oxide board, chip board ceilings are some of the example of panel absorbers. These panels are used for ceiling or wall acoustic applications. These panels can also be used for decoration purposes. Since panel does not have their own acoustic properties, they are used along with acoustic absorber material. For example in roof ceiling application, the ceiling panel is used in conjunction with an acoustic absorber in order to have desired acoustic properties. Most of the acoustic absorber materials were already discussed in Sects. 1 and 2. At times, in order to achieve good acoustic properties, the acoustic absorber materials are placed at certain distances (or air gap) from the panels.

Acoustic properties of a perforated wooden panel backed by PET/cotton/wool (70/30) needle-punched mats have been studied by Seddeq et al. [12]. The mats were placed at 1, 2, 3 and 4 cm air gap behind the panel. The wooden panel used had a 6 % perforation ratio. As the air gap increased, the NAC value increased particularly for the low to mid frequencies (100–1600 Hz), with a maximum value of 0.90 obtained at 1100 Hz for the sample with 4 cm air gap. In this study they used a single layered of PET/cotton/wool needle-punched mat. When the mat layered was increased to 2 and 3, the NAC curve shifted towards the low frequency absorption.

Zulkarnain and Nor [22] studied the acoustic properties of a coir fibre based material. Coir fibres were bonded with latex and were backed by a woven cloth. A perforated metal plate with 0.20 perforation ratio, 1 mm plate thickness and 2 mm perforated hole diameter along with air gap backing was used in the study. The length of the air gap was not mentioned in this study. By comparing the NAC values of 20 mm thickness latex bonded coir mat with and without perforated panels, perforated panel showed better acoustic values in the frequency range of 600–2400 Hz. After that, this value slightly decreased.

In another study, Fouladi et al. [23], studied the acoustic properties of a multi-layer acoustic panel consisting of coir fibre mat (latex bonded), air gap and perforated panel/plate for wall insulation application. The main aim behind this study was to improve the acoustic properties, particularly in the low frequency ranges without having a large air gap, while placing the multi-layered assembly. They used two differ types of arrangements of perforated plate (made of aluminium alloy), coir fibre mat and air gap assembly. In one arrangement, perforated plate, followed by coir fibre mat and air gap were used, while in another arrangement, coir fibre mat, perforated plate and air gap were used. In both case, 1 and 20 %

perforation levels of the plate were used in the study. NAC curve of the perforated plate having 20 % porosity, showed acoustic properties which can absorb wide range of frequencies from 0 to 5000 Hz. Whereas, NAC curve of the perforated plate having 1 % porosity showed that it can mainly absorb low frequency (0–1000 Hz), after that the absorption tendency decreased. In both cases, perforated plates were backed with 50 mm coir fibre layer and 35 mm air gap. When the plate perforation level was at 1 %, it acts like a deflector for sound waves, thereby resulting in gradually diminishing acoustic properties. Authors also repeated the same study for 50 mm coir fibre layer backed by perforated plates and 35 mm air gap. In that case, for both 1 and 20 % perforation levels, there was not much difference between the NAC curves, with high absorptions of low frequencies for both perforation levels. These results suggest that placing of various layers also affects acoustic properties. In another case, authors obtained similar NAC curves for 50 mm coir fibre layer backed with 80 mm air gap and 50 mm coir fibre layer backed with perforated plate and 35 mm air gap. Authors showed that it was possible to absorb most of the frequencies without going for a large air gap, by the proper selection and position of a layered assembly.

5 Acoustic Materials Coated, Sprayed or Embedded with Fillers

In order to improve the performance of acoustic materials in various applications, sometimes materials were coated or bonded with various chemicals. This bonding can also be in the form of using low melt PET, PP or, bi-component fibres, which can be bonded at thermal bonding method of manufacturing. Some of this aspect has been already discussed in previous sections [18, 10, 11]. Furthermore, acoustic materials are also covered with microporous film and aluminium foil, coated or sprayed with flame retardant and silicone in order to improve fire, thermal insulation, moisture resistance properties amongst other properties.

In one study, Patnaik et al. [3] studied the acoustic properties of building materials made from waste wool and RPET fibres. Since material requirements for building industries are stringent in terms of fire properties, needle-punched mats were sprayed with a low level of fire retardant (5 % by weight, comprising a mixture of diammonium phosphate and sodium tetraborate) to impart fire retardancy properties. This treatment also assists in eliminating fungus and moth problems associated with the waste wool fibres. Furthermore, 1 % silicon (by weight) was sprayed on the waste wool fibres in order to improve their moisture resistance. All spraying was carried out on the surface layers only in order to maintain the recyclability and biodegradability of the sample. The NAC values of these samples were in the range of 0.61–0.75 in the overall frequency range of 50–5700 Hz. The above chemical spraying did not change the NAC values of the samples as the spraying was done sparingly at the surface level only. Since it was spraying as opposed to coating, most of the pores were available for sound absorption application.

Venkataraman et al. [24] studied the acoustic properties of amorphous silica aerogel embedded (PP/PET) nonwovens in the frequency range of 50–6400 Hz. They used three different nonwovens having thicknesses 3.2–6.6 mm and area weights 272–499 g/m^2, respectively. The NAC values were low at low frequencies (50–1600 Hz), thereafter gradually increasing for high frequencies (2000–6400 Hz). A maximum NAC value of 0.90 was obtained at 6200 Hz for the sample with higher area weight and thickness as compared to thinner and light weight sample (3.2 mm thickness and 272 g/m^2) which has a NAC value of 0.40 at that frequency.

Seddeq [25] studied the acoustic properties of nonwoven mat covered with polyvinyl chloride (PVC) film and compared with the results of the mat without film in the frequency range of 250–5000 Hz. Authors reported that PVC attachment increased sound absorption at low and mid frequencies at the expense of higher frequencies.

In a patent, Horton [26] discussed about the development of fire retardant thermal and acoustic insulation material. A nonwoven mat was prepared from recycled and shoddy materials. A layer of aluminium foil was bonded to one side or both side of the mat in order to create a product which reflects heat as well as and high frequency sound waves. At the same time, the above material was capable of absorbing low frequency sounds waves. This material can be potentially used for automotive or aircraft sound reduction applications. Furthermore, this mat was coated with a fire retardant in order to improve fire properties.

6 Use of Nanomaterials

Nanomaterials in the form of nanofibre mats have been used to improve the acoustic properties of the conventional materials. Some of the advantages of using these materials are their ability to absorb low and medium frequencies, which are difficult to absorb in traditional acoustic materials. In order to absorb them, conventional acoustic material weight and thickness has to be increased. With the use of nanofibre mats, the weight and thickness issues of the conventional acoustic material have been addressed. The nanofibres were used as layers on a fabric or sandwiched in between two fabrics for acoustic applications. Nanofibre mats produced from electrospinning, melt blowing, melt spinning and other technologies have been used to improve the acoustic properties of the materials [13, 27, 28, 29, 30, 31].

Na et al. [29] measured the NAC values of multi-layered nanofibres web produced from melt spinning polyamide (Nylon) and compared it with the NAC values of the microfibre-based materials. The average diameters of nanofibres obtained from above process were 800 nm. The area weight of the nanofibre layer was 30 g/m^2. Several layers of the nanofibre web were stacked over one another in order to match with the area weight of the knitted microfibre fabrics. The area weight of the nanofibre fabrics were varied from 180 to 540 g/m^2, having 6–18 nanofibre layers in order to match with the area weight of the knitted microfibre fabrics. The NAC results showed that nanofibre fabrics were absorbing more noise as compared to the microfibre fabrics in the frequency range of 1000–4000 Hz. Furthermore, with increasing number of nanofibre layers, NAC values were also increased. The reason for such increase in the NAC values may be due to large surface area of the nanofibre fabrics and presence of very fine pores which increases frictional contact of the sound wave with the pores, thereby converting sound energy into thermal energy [29]. Another possible reason for this may be certain frequencies vibrate the nanofibre layer into those frequencies, thereby creating a resonance frequency, and dampening the sound wave [13, 27, 30]. There were not much change in the NAC values in the low frequency range of 50–1000 Hz for both nanofibre and microfibre fabrics [29].

Na et al. [29] also studied the influence of nanofibre layers on a knitted fabric (1187 g/m^2) and compared with the NAC values obtained from the similar fabric without any nanofibre layers (1000 g/m^2). From 125 to 1000 Hz, there was not much change in the NAC values of both the fabrics. A gradual increase in NAC values were observed for both fabrics in frequency range of 2000–4000 Hz, with around 85 % increase in NAC value was observed at 4000 Hz.

Polyvinyl alcohol (PVA) nanofibre web/membrane produced by electrospinning process was evaluated for their NAC values [27]. The average diameters of nanofibres obtained were 280 $\pm$ 80 nm. The area weight of the nanofibre web was 17 g/m^2. Water vapour was applied to the surface of nanofibres in a varying time interval from 10 to 120 s in order to change the structure of the nanofibres. With the increasing water vapour time, the NAC values of the nanofibre webs were increased up to 2800 Hz, after than it were decreased. The possible reason for this was that PVA was water soluble, with increasing water vapour content, the irregularity in the nanofibre web were increased and after a limit, majority of fibres were merged in water vapour, thereby decreasing NAC values. In the low frequency range of 100–1000 Hz, NAC values were 0.15–0.18 for water vapour time intervals 30, 90 and 120 s, respectively.

Acoustic properties of Nylon 6 nanofibre produced by electrospinning process were studied by Trematerra et al. [31]. The nanofibre layers were glued to different porous material surfaces (foam, kenaf and PET nonwoven mats) having different thickness. The nanofibre diameters were in the range of 150–200 nm. The thickness of the nanofibre layer was 10 μm. By comparing the NAC values of the kenaf mat (40 mm thickness) with and without nanofibre layers, there was increase in this value for nanofibre layer glued mat for 200–2000 Hz. At 500 Hz, the NAC values were 0.35 and 0.25, respectively, for with and without nanofibre layers. Similarly, at 1300 Hz, these values were, 0.80 and 0.70, respectively. These results showed that addition of nanofibres layers marginally increased the NAC values in the low–medium frequency range. In another experiment authors used PET nonwoven mat having 3 and 4 mm thickness and compared with the NAC values of with and without nanofibre layers in 200–6300 Hz [31]. There were no changes in the NAC values up to 2000 Hz having nanofibre layers irrespective of mat thickness. There was a marginal increase in NAC values for 2000–6300 Hz for mats having a nanofibre layer.

Asmatulu et al. [32] studied the acoustic properties of polyvinyl chloride (PVC) nanofibres produced by varying electrospinning process parameters. These nanofibres were produced for reducing aircraft interior noise application. The PVC nanofibre diameters were in the range of 200–900 nm. The NAC values of PVC nanofibres were compared with melamine foam samples having similar weights as that of nanofibres. The weights of the PVC nanofibres were 0.5, 1 and 1.5 gm, respectively. The weight of the melamine foam samples were 2 and 4 gm, respectively. There were no significant changes in NAC values of both set of were observed for low frequency 50–1000 Hz. But for 2000–6200 Hz, NAC values of the nanofibres were higher than that of the melamine foam samples. The 1.5 gm PVC nanofibres samples showed higher NAC values as compared to 2 gm melamine foam sample.

Alba et al. [33] studied the acoustic properties of nanofibres used along with a PET mat. In this study, authors have not specified about the type of nanofibres used and their diameters values. In one of the experiment, they found that NAC values for PET mat with a nanofibre layer increased in the low frequency 125–1000 Hz. A maximum increased was observed around 500 Hz, where the NAC values were 0.40 for a PET mat with a nanofibre layer and 0.20 for a PET mat without nanofibre layer. While comparing other frequency range, from 1000 to 2500 Hz, a further increase in the NAC values were observed for the PET mat with a nanofibre layer. At 1600 Hz, the NAC values were 0.92 for a PET mat with a nanofibre layer and 0.80 for a PET mat without nanofibre layer. Authors also studied the NAC values of various samples, which include drilled panel, drilled panel with a polyester mat, drilled panel with a polyester mat along with nanolayer and drilled panel with mineral wool [33]. The NAC values of drilled panel with a PET mat along with nanolayer was the highest (0.95) among the layers in the low frequency 125–1000 Hz. From 1000 to 5000 Hz, there was not much of a difference between the samples, with a gradual reduction in NAC values with the increase in the frequency.

Polyacrylonitrile (PAN) and PU nanofibre layers produced by electrospinning process were evaluated for their STL and NAC values [34]. The average diameters of PAN and PU nanofibres obtained were 121 and 203 nm, respectively. The area weights of the nanofibre layers were 1, 3 and 5 g/m^2. These layers were sandwiched between two nonwoven layers (PET and wool) in order to obtained final samples. As compared to the PET and wool mats, the STL values of the PET and wool mats with nanofibre layer showed an increase in the overall frequency of 50–6200 Hz. There was an increase STL values in the low frequency of 50–1000 Hz, followed by subsequent increase from 2000 to 6200 Hz. With the increasing number of nanofibre layers (1–5) resulted in higher STL values. PAN based sampled showed higher STL as compared to PU samples because of presence of smaller nanofibres. The NAC values of PET mat increased from 0.40 to 0.67 and 0.71 by the presence of PAN and PU nanofibres layers, at a frequency of 2000 Hz.

Some of the drawbacks of using nanomaterials are the low production rate of nano web, requirement of high voltage or similar sources, solvent recovery problem, poor mechanical strength and handling problem of the nano web.

7 Recent Trends and Future Directions

The approach of active noise control is becoming popular nowadays, which work in combination with the passive noise control for developing hybrid sound absorbers. The active noise control approach is particularly effective in controlling the low frequency noise. Therefore, in order to achieve a good noise control over a wide range of frequencies, hybrid absorbers were used. Figure 3 shows the hybrid absorbers consisting of a passive porous structure and an active controller at its rear side, which can be digitally controlled [13].

The design of broadband sound absorber or smart foam which works on the principle of hybrid sound absorbers is also the focus of recent research [35]. This absorber was prepared from melamine foam with polyvinylidene fluoride (PVDF) piezoelectric film embedded actuators. The piezoelectric actuators work as the active noise control whereas the porous wire meshes work as the passive noise control. The low frequency sound is controlled by the active cancellation process and the medium to high frequency sound is absorbed by the melamine foam's passive properties.

Micro-perforated panels are also another research area, which work in the principle of combined active–passive noise control. The use of perforated panel absorbers can be date back to many years and some of them were discussed in Sect. 4. Recently, these panel absorbers are prepared with pore dimensions in the sub-millimetre region [36]. When the sound waves travel in a micro-perforated panel, the acoustic energy is converted into heat due to the friction between the internal surface of the perforation and the sound wave travelling in the air. The perforation can be in the form of holes or slots with diameters less than 0.3 mm. These perforations produce sufficient resistance to sound wave to produce wide range of sound absorption. Reducing the diameter below this, does not provide much benefit for wider absorption bandwidth.

The use of nonmaterial in acoustic absorption is also a major area of research by several researchers. Section 6 describes the scope of using nanomaterial. The advantages of using nanomaterials are their ability to absorb low and medium frequencies at lower material weight and thickness, as compared to the traditional thicker acoustic materials. Several challenges such as low production rate, strength and handling issues of the nanoweb and solvent recovery problem need to be resolved amongst others before the nanomaterials are successfully used in acoustic applications.

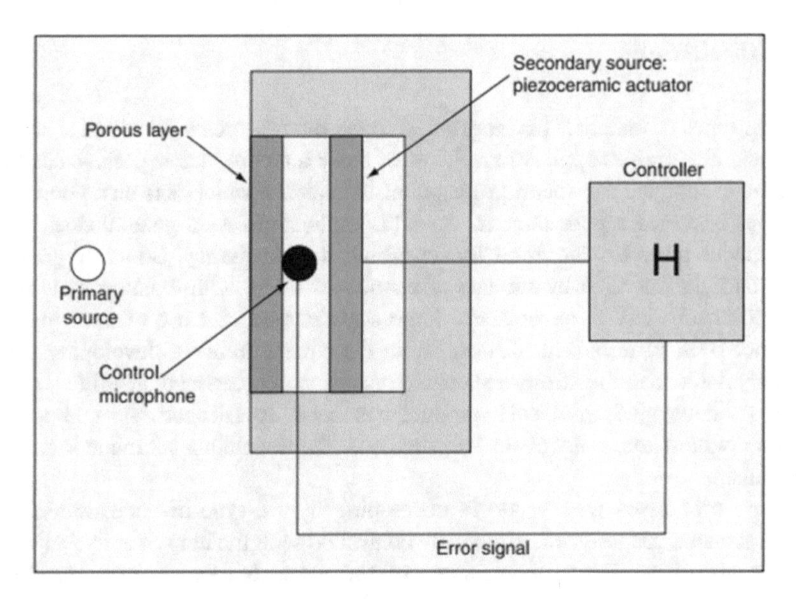

Fig. 3 Hybrid passive/active absorber cell (*Source* Arenas and Crocker [13])

A study by Jiang et al. [37] evaluated activated carbon material for the use as high performance acoustic absorbers. The study established that the activated carbon fibre composites exhibited higher capacity to absorb normal incidence sound waves as compared to the composites with either glass or cotton fibres. The activated carbon material can be prepared from carbonaceous materials by the process of carbonization and activation. Carbonization is a thermomechanical process (pyrolysis) that converts carbonaceous materials (such as coconut, wood, bagasse and shells: organic; and synthetic resins) into active carbon products. The major precursors used for preparing activated carbon fibres are PAN, Novoloid (Novolac resin), rayon and acrylic. Viscose rayon was mainly used for the activated carbon fibre preparation for the last few decades.

The acoustic absorption depends on the thickness of the panel as well as the thickness of the fibrous absorber. Hence, it is a real challenge to design thin sound absorbers with enough sound absorption. However, the development of finer machineries and advanced technologies such as welded meshing, powder metallurgy, laser drilling and electroetching, can help in achieving this by incorporating holes with micrometre diameter. Modelling techniques such as Finite element analysis (FEA), computational fluid dynamics, (CFD) can be also used to for effective design and property optimization of acoustic absorbers. Other future directions and research trends are discussed in Chapter "Design of Acoustic Textiles: Environmental Challenges and Opportunities for Future Direction".

8 Conclusion

Various types of materials have been developed over the years for sound absorption and transmission loss applications. Most of these materials act as passive absorbers in order to mitigate the sound propagation. A material which acts as a good sound absorber becomes a poor material for STL application. As a general rule, thicker and heavier materials are good for absorbing low frequency. Some progress has been made in this area by the use of nanofibres along with existing materials in order to absorb low frequency. But large scale commercial use of such materials have not been materialized till now. With the current trend of developing natural and recyclable materials from sustainable resources, increasing demand for acoustic materials developed from such product has been experienced. As compared to woven counterparts, nonwovens still dominate the developments made in acoustic materials.

Some progresses were made in developing hybrid type of acoustic absorbers where acoustic property was effectively controlled with the help of active as well as passive absorbers. Furthermore, some of the research focus is now developing thinner and lighter fibrous materials apart from the nanomaterials. Also, there is ongoing research work in developing panel absorbers having micrometre diameter holes in the panel.

Acknowledgments The author acknowledged the following sources for granting permission to reproduce some figures and tables used in this chapter.

Reprinted from Patnaik et al. [3], Copyright (2015), Arenas and Crocker [13] with permission from Elsevier, CSIR [1] and Sound & Vibration.

References

1. CSIR (2015) Developing acoustic materials. CSIR report, Port Elizabeth, South Africa
2. Mvubu M, Patnaik A, Anandjiwala RD (2015) Optimization of process parameters of sound barrier application. J Eng Fibres Fabr 10(4):47–54
3. Patnaik A, Mvubu M, Muniyasamy S, Botha A, Anandjiwala RD (2015) Thermal and sound insulation materials from waste wool and recycled polyester fibres and their biodegradation studies. Energy Build 92(1):161–169
4. Nayak R, Padhye R, Fergusson S (2012) Identification of natural textile fibres. In: Handbook of natural fibres, vol 1. Types, properties and factors affecting breeding and cultivation. Woodhead Publishing Ltd, UK, p 314
5. Thilagavathi G, Pradeep E, Kannaian T, Sasikala L (2010) Development of natural fibre nonwovens for application as car interiors noise control. J Ind Text 39(3):267–278
6. Parikh DV, Chen Y, Sun L (2006) Reducing automotive interior noise with natural fibre nonwoven floor covering systems. Text Res J 76(11):813–820
7. Oldham DJ, Egan CA, Cookson RD (2011) Sustainable acoustic absorbers from the biomass. Appl Acoust 72(6):350–363
8. Chen Y, Jiang N (2007) Carbonized and activated nonwovens as high performance acoustic materials: Part I noise absorption. Text Res J 77(10):785–791

9. Nayak R, Padhye R (2014) Introduction: the apparel industry. In: Nayak R, Padhye R (eds) Garment manufacturing technology. Elsevier, Amsterdam, pp 1–18
10. Lee Y, Joo C (2003) Sound absorption properties of recycled polyester fibrous assembly absorbers. AUTEX Res J 3(2):78–84
11. Manning J, Panneton R (2013) Acoustical model for shoddy-based fibre sound absorbers. Text Res J 83(13):1356–1370
12. Seddeq HM, Aly NM, Marwa AA, Elshakankery MH (2012) Investigation on sound absorption properties for recycled fibrous materials. J Ind Text 43(1):56–73
13. Arenas JP, Crocker MJ (2010) Recent trends in porous sound-absorbing materials. Sound Vibra 44(7):12–17
14. Shoshani YZ (1990) Effect of nonwoven backings on the noise absorption capacity of tufted carpets. Text Res J 60(8):452–456
15. Shoshani YZ, Wilding MA (1991) Effect of pile parameters on the noise absorption capacity of tufted carpets. Text Res J 61(12):736–742
16. Heylighen A, Rychtáriková M, Vermeir G (2010) Designing spaces for every listener. Univ Access Info Soc 9(3):283–292
17. Moholkar VS, Warmoeskerken MMCG (2003) Acoustical characteristics of textile materials. Text Res J 76(11):813–820
18. Kucuk M, Korkmaz Y (2012) The effect of physical parameters on sound absorption properties of natural fibre mixed nonwoven composites. Text Res J 82(20):2043–2053
19. Yang S, Yu WD (2011) Air permeability and acoustic absorbing behaviour of nonwovens. J Fibre Bioeng Infor 3(4):204–208
20. Suvari F, Ulcay Y, Maze B, Pourdeyhimi B (2013) Acoustical absorptive properties of spunbonded nonwovens made from islands-in-the-sea bicomponent filaments. J Text Inst 104 (4):438–445
21. Bo Z, Tianning C (2009) Calculation of sound absorption characteristics of porous sintered fiber metal. Appl Acoust 70(2):337–346
22. Zulkarnain RZ, Nor MJM (2010) Noise control using coconut coir fibre sound absorber with porous layer backing and perforated panel. Am J Appl Sci 7(2):260–264
23. Fouladi MH, Nor MJM, Ayub M, Leman ZA (2010) Utilization of coir fibre in multilayer acoustic absorption panel. Appl Acoust 71(3):241–249
24. Venkataraman M, Mishra R, Arumugam V, Jamshaid H, Militky J (2014) Acoustic properties of aerogel embedded nonwoven fabrics. In: Proceedings of the Nanocon 2014, 5–7 Nov 2014, Brno, Czech Republic, pp 1–8
25. Seddeq HM (2009) Factors influencing acoustic performance of sound absorptive materials. Aust J Basic Appl Sci 3(4):4610–4617
26. Horton BD (1999) Fire hydrant thermal and acoustic insulation material. US patent Publication number US5955386 A, 21 Sept 1999
27. Kalinová K (2011) Nanofibrous resonant membrane for acoustic applications. J Nanomater 1–6
28. Mohrova J, Kalinová K (2012) Different structures of PVA nanofibrous membrane for sound absorption application. J Nanomater 1–4
29. Na Y, Agnhage T, Cho G (2012) Sound absorption of multiple layers of nanofibre webs and the comparison of measuring methods for sound absorption coefficients. Fibres Polym 13(10):1348–1352
30. Tascan M, Vaughn EA (2008) Effects of fibre denier, fibre cross-sectional shape and fabric density on acoustical behaviour of vertically lapped nonwoven fabrics. J Eng Fibres Fabr 3(2):32–38
31. Trematerra A, Iannace G, Nesti S, Fatarella E, Peruzzi F (2014) Acoustic properties of nanofibres. In: Forum acusticum, 7–12 Sept 2014, Kraków, Poland, pp 1–4
32. Asmatulu R, Khan W, Yildirim MB (2009) Acoustical properties of electrospun nanofibres for aircraft interior noise reduction. In: Proceedings of the ASME 2009 international mechanical engineering congress and exposition, IMECE2009, 13–19 Nov, Lake Buena Vista, Florida, USA, pp 1–5

33. Alba J, Rey RD, Berto L, Hervás C (2012) Use of textile nanofibres to improve the sound absorption coefficient of drilled panels for acoustic applications. In: Proceedings of the acoustics 2012 Nantes conference, 23–27 April 2012, Nantes, France, 303–307
34. Rabbi A, Bahrambeygi H, Shoushtari AM, Nasouri K (2013) Incorporation of nanofibre layers in nonwoven materials for improving their acoustic properties. J Eng Fibres Fabr 8(4):36–41
35. Galland MA, Mazeaud B, Sellen N (2005) Hybrid passive/active absorbers for flow ducts. Appl Acoust 66(6):691–708
36. Asdrubali F, Pispola G (2007) Properties of transparent sound absorbing panels for use in noise barriers. J Acoust Soc Am 121(1):214–221
37. Jiang N, Chen JY, Parikh DV (2009) Acoustical evaluation of carbonized and activated cotton nonwovens. Bioresour Technol 100(24):6533–6536

Manufacturing Methods for Acoustic Textiles

Vikas K. Singh and Samrat Mukhopadhyay

Abstract Research community is looking for non-conventional approaches and materials to minimize the noise pollution considering its long-term ill effects on human health. Textile materials were studied earlier for their acoustic properties but as the menace of noise pollution has become more prominent in recent decades, textiles are looked upon as a cheaper, simpler and effective alternative to existing materials and approaches. This chapter focuses on the manufacturing techniques used to produce various products of textile materials finding application in acoustic insulation. Fibrous mats, needlepunched nonwovens and polymers composites made of nonwovens are among widely researched textile materials nowadays. Some processes produce dense textile structures that reflect more sound waves than it absorbs while some processes tend to produce more porous textile structures that absorb sound waves by dissipating their energy while they pass through tortuous path. The latter approach is widely used because most of the textile materials are inherently porous and it is expected that this property will lead to more sound absorption. Till date, various methods have been reported to produce thick textile structures with sufficient amount of porosity. The approach of combining two or more textile structures is also used to counter the noises at low frequencies, which are not usually blocked by textile materials. Final part of the chapter deals with emerging techniques that are reported by very few researchers and may lead to more interesting outcomes in upcoming years.

Keywords Sound · Acoustic barrier · Noise pollution · Nonwoven · Textile · Composite · Insulation · Passive noise control

V.K. Singh (✉) · S. Mukhopadhyay
Department of Textile Technology, IIT Delhi, New Delhi 110016, India
e-mail: vikas.texogenic@gmail.com

R. Padhye and R. Nayak (eds.), *Acoustic Textiles*, Textile Science and Clothing Technology, DOI 10.1007/978-981-10-1476-5_5

1 Introduction

With the profound use of machinery in industries, transportation and day-to-day usage, noise pollution has become a major cause of discomfort to the mankind. Knowingly or unknowingly, we come across low to high amount of noise pollution everyday for a considerable amount of time and it is usually an underrated type of pollution that has drawn least attention as compared to water or air pollution. Most of the workplaces are generally in annoying environment of prolonged and harmful noise levels. Noise not only has an adverse effects on physical and psychological health, but also on the rate of ageing of man-made structures [1]. Various efforts are made for a long time to tackle the menace of noise pollution. Noise can be diminished by active or passive means. Sound is a wave and active noise control makes use of the destructive interference phenomena. In active noise control, the sound wave produced by the sound source is cancelled by a secondary sound source [2]. Though effective, this technique is not so popular in usual day-to-day life but it deals effectively with low frequency sound. On the other hand, passive noise control works by protecting the field from the sound wave. It can be done by reducing the noise at the source itself by checking any unnecessary vibrating surface, fluid jet stream, etc. Another way is by blocking the path by which the sound wave travels. The final way is to protect the receivers end, i.e. human ear in most cases, from the undesirable sound waves. This way is adopted when all other means are not sufficient or are ineffective. It also blocks other sound waves that are not disturbing for us. Overall, the passive control at sound source or receivers end is usually not very practical in many cases. The only option left is to block the unwanted sound waves in between.

The transmission of sound waves is reduced by placing a sound insulating material or a sound absorbing medium in its path. The sound insulating materials are usually dense materials that reflect back the sound waves which incident on it, whereas the sound absorbing materials absorb the sound waves. Sound absorbing materials are usually porous and thus most of the textile materials can be used for this purpose as they are inherently porous in nature [3]. Heavy textile structures in the form of dense fabrics or multi-layered structures usually in combination with other structures are being studied for sound insulation also. In order to overcome the low performance of textile materials in sound insulation and use their advantage of low cost and customizability, a combination of two or more materials and approaches is considered. The research in textile-based acoustic insulation targets mainly on optimizing various customizable parameters rather than inventing new production technique to achieve best sound barrier properties.

Textile materials provide the advantage of low production cost and freedom to customize end properties. It is due to these characteristics, the textile materials are explored for sound absorption and insulation purposes [4]. In most cases, textile materials are not manufactured primarily for acoustic insulation, as this is a secondary desired property and therefore the research community emphasizes more on understanding of process and material parameters that affect sound insulation.

Mostly, the efforts are made to optimize various factors in the construction of textile material to give higher sound barrier properties. There is limited literature dedicated towards production of textiles exclusively for sound insulation. The manufacturing techniques also do not differ too much from the conventional methods as the idea is to use the existing textiles for the purpose. There are different techniques that aim towards production of different types of fibres, fabrics and other structures that resemble the conventional production techniques. However, understanding of important parameters of textile structure such as material, yarn count, yarn twist, air permeability, etc. are crucial in their production.

The available literature mostly includes the use of nonwovens. Needlepunching is the most popular technique that has been reported to produce nonwovens. Second most explored are the polymeric composites based on textile materials. Comprehensive studies on various types of fibres, fibre blends, carpets, 3D fabrics and other textile structures have also been carried out that will be discussed in this chapter.

2 Fibres as Raw Material

The basic raw material for textile structures is fibre. Therefore, using fibres that are inherently sound insulating will ultimately help in production of textile materials for sound insulation. Both natural and synthetic fibres are studied. Synthetic fibres provide a large window of customizing the cross section, fineness and thus the overall shape factor and surface area. This is not usually true with the natural fibres because there is virtually no control over production of these fibres. Natural fibres, usually lignocellulosic fibres obtained from plants and trees, are considered upon a cheap alternative because they are abundantly available in nature and do not cause environmental concerns at the time of disposal. Surface modification techniques have been found to improve the sound barrier properties. Hemp fibres showed higher sound absorption when higher concentration of alkali, up to 15 % w/v, was used for treatment. It was attributed to a combined effect of weight loss and air flow resistivity of nonwoven hemp fabrics treated at higher alkali concentrations [5]. Use of polyethylene glycol and glycerol improves the transmission loss in cellulose-based composites [6].

Hollow fibres are considered to absorb more sound waves due to the air cavity present in them. Figure 1 shows the microscopic image of hollow Estabragh fibres.

Kapok fibres and Estabragh fibres were used in production of nonwovens with polyester and polypropylene fibres, respectively. Addition of these hollow fibres led to higher sound absorption than the ones made of only polyester or polypropylene fibres [7, 8]. Other natural lignocellulosic fibres such as coir, flax, typha, etc. have also been used as a raw material.

Synthetic fibres, especially the thermoplastic fibres, have been investigated in terms of cross-sectional shape and fineness. It is observed that finer fibres result in better sound barrier properties and it is attributed to larger surface area of fibres and more tortuous path in nonwovens for sound wave to travel [9–11]. Trilobal and

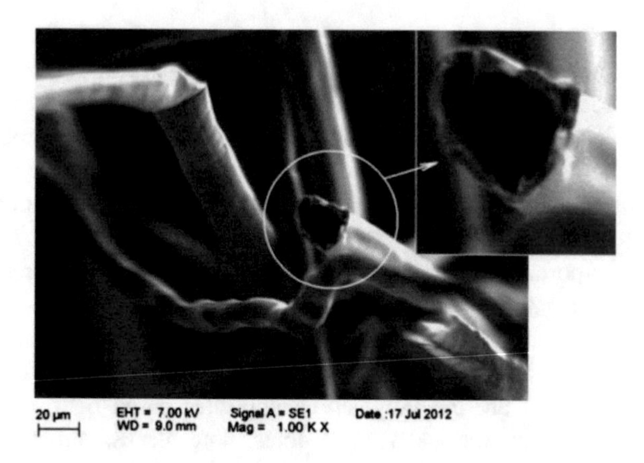

Fig. 1 Cross section of Estabragh fibre (adapted from Hassanzadeh et al. [8])

4DG polyester fibres showed better sound insulation properties in nonwovens as compared to the round cross-section fibres of same denier [12]. An invention discusses in detail the use of regularly and irregularly profiled cross section and contour of fibres. It was claimed that polyethylene and polypropylene fibres having such cross section and contour achieve greater sound damping when used in knitted fabric roll for compressed air devices [13]. Fibres with different types of cross sections and contours as reported in a patent are illustrated in Fig. 2.

Besides melt spinning, finer fibres are produced by electrospinning also [14]. However, the fibres spun from electrospinning are extremely fine and the dimensions of diameter may vary in the range of few nanometres to micrometres. Production of fibres from electrospinning on a large scale for sound barrier applications is not practically feasible. Although there are some inventions that refer to layered electrospun and nonwoven webs the electrospun fibres are yet to be commercialized for acoustic applications. A layered nonwoven fabric with an acoustic damping membrane made up of electrospun nanofibre web in between the carded web performed better even in low frequency noise [15].

Suvari et al. [4] used islands-in-the-sea bicomponent filaments where island was Nylon-6 polymer and sea was polypropylene polymer. These filaments were used in nonwoven by spunbonding and hydroentangled to fibrillate the fibres. Repeated hydroentangling led to higher fibrillation as can be seen in Fig. 3. It was observed that the fibrillation of filaments to finer fibres led to higher sound absorption coefficient in the higher frequency region.

3 Woven and Nonwoven Fabrics

It is assumed that the sound waves enter into the porous and tortuous path of the fabrics and is converted to heat by internal friction. Fabrics made up of microfibre yarns provide more fibre surface and tortuous path than fabrics with regular fibre

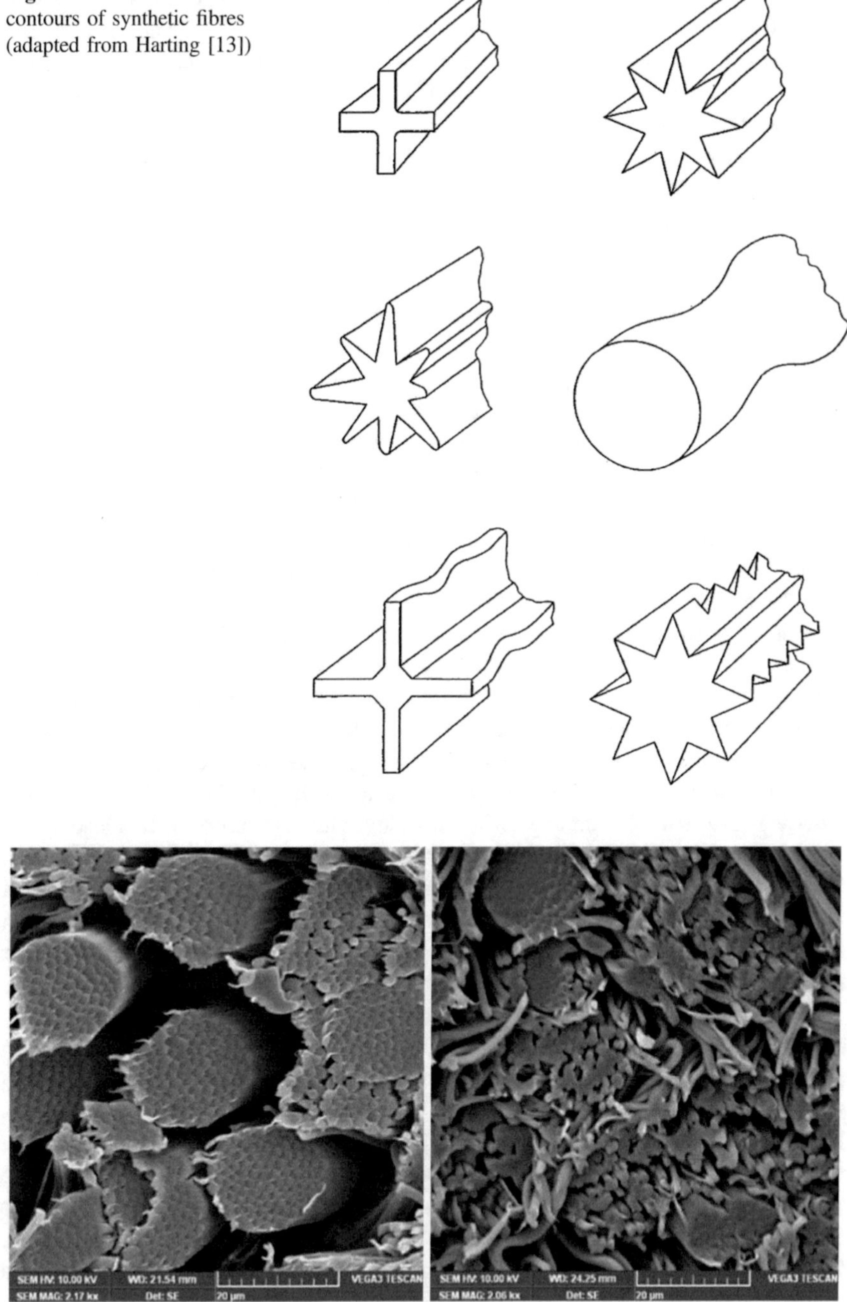

Fig. 2 Cross section and contours of synthetic fibres (adapted from Harting [13])

Fig. 3 Cross section of nonwoven web after one and three passes (adapted from Suvari et al. [4])

yarns. The sound absorption property of these fabrics is higher at both lower and higher frequencies. Usually the woven fabrics are not very good candidate for use in sound barrier applications. They have a thin structure that is not dense enough to reflect the sound waves nor have a highly porous structure to provide enough tortuous path to dissipate the sound energy. Very few comprehensive studies have been reported that compare effect of different weave patterns and weaving parameters for sound barrier properties. Air permeability is often related to the noise reduction of woven fabrics and a study carried out by Soltani and Zarrebini [3] discussed its role using woven fabrics with different pick density as well as weft yarn twist. Same research group reported detailed study of effect of different weave patterns, yarn spinning system, back air cavity and weft density on sound absorption of woven fabrics [16]. On the other hand, nonwoven fabrics are widely studied for sound barrier properties. The inherent porous nature of nonwovens makes them a better choice over the woven fabrics. The ease of manufacturing, versatility in choice of fibres and less time consumption are among the few advantages of nonwoven fabrics. The studies on sound barrier properties use nonwovens as such, as layered structures, as reinforcement in polymeric composites, in combination with woven fabrics, etc.

Needlepunching technique is the most popular process that is studied for making nonwovens for sound insulation purpose. In this technique, the fibres in the web are mechanically entangled by the repeated penetration of specially designed barbed needles. The fibres or mixture of fibres is subjected to opening and web formation in carding process before it is fed to the needlepunching operation to entangle the fibres. There are various factors that are varied during the needlepunching operation such as web orientation, final weight per unit area, needlepunching density, mass fraction of various fibres, etc. to get the desired end product. Needlepunching process is fast and efficient even when working with natural fibres. It is also useful to make nonwovens of brittle natural fibres by mixing with conventionally used fibres that are otherwise difficult to process. A study used nonwoven mats of viscose fibres as raw material and processed them to make felts of activated carbon fibres. Light weight, anti-corrosion and fireproof carbon fibre felts appear better sound absorbing materials than other conventional nonwovens [5].

Besides needlepunching, works are also reported on nonwovens produced by spunbonding and water lay techniques. As mentioned earlier, the study of Suvari et al. [4] comprised of island-in-the-sea bicomponent filaments. These fibres consisted of 75 % of 'Island' polymer, i.e. Nylon-6 and 25 % of 'Sea' polymer, i.e. Polypropylene. Nonwoven webs with different island counts were produced to analyze their effect on sound absorption property. The spunbonded webs were bonded by hydroentangling. Hydroentangling also resulted in fibrillation of the filament giving rise to exposed finer Nylon fibres. As compared to electrospinning, this process is more efficient in bulk production and comparable in sound absorptive properties. A setup for hydroentangling of the fibrous web is shown in Fig. 4.

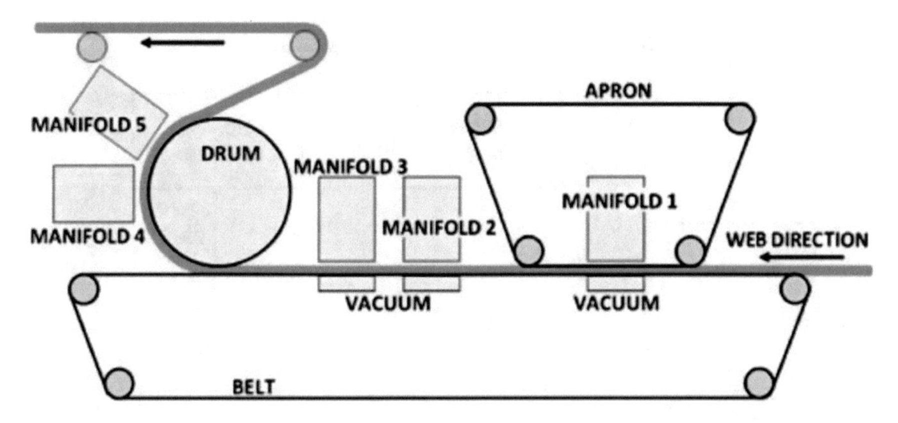

Fig. 4 The hydroentangling of island-in-the-sea spunbonded fibres (adapted from Suvari et al. [4])

The wet laid technique is preferred for much smaller fibre dimensions and pulp of lignocellulosic origin. The slurry is made to settle on a perforated surface and then compressed to make sheets. One such work is reported by Fages et al. [17] where nonwovens from flax fibres and Polyvinyl alcohol (PVA) fibres are made into sheets by wet laid technique. Initially the fibres were weighed to predetermined amount and individualized using a high shear pulping machine. The concentration of fibres in aqueous dispersion was maintained at very low amount of 10 g/L. This dispersion was further diluted to 1 g/L and was slowly stirred throughout the process to avoid agglomeration of fibres. Right before dropping the prepared dispersion on the perforated collection strip, it was further diluted to 0.33 g/L. These sheets were passed through hot air for drying and later calendered by hot rollers. Surface temperature of rollers was maintained at 200 °C. Resulting nonwoven structure was a thermobonded sheet of flax fibres and PVA and tested for sound absorption properties. Although wet laid process uses large amount of water, it is considered to be eco-friendly because the water can be recovered in the hydroformer station. Passage of dispersion and recovery of unused water can be seen in Fig. 5.

Another study focusing on using powder-like waste generated from shredded waste of textiles in order to use it for acoustic insulation by wet laid process is reported. The research team used recycled polyethylene terephthalate (PET) waste from bottles in the mixture for thermally binding the structure [18].

A different kind of thermal bonding of air laid web of two different compositions of polyester, cotton and bicomponent fibres was reported recently. Figure 6 includes the final sample and the setup for preparation of samples. It can be seen that the web was placed under compression and heat was provided from one side only. This process resulted in nonwoven structures that had a density gradient. The surface in contact of the heated plate undergone effective thermal bonding and thus was denser than the other surface. Samples were compared for sound absorption coefficient with both the faces alternatively towards the sound source. The samples facing the sound source with denser face showed less sound absorption because most of the sound

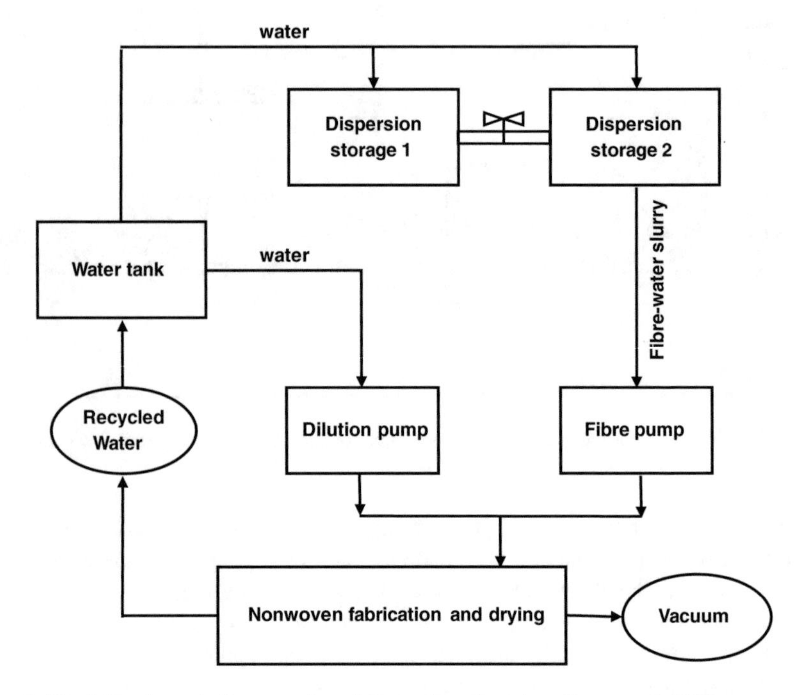

Fig. 5 Wet laid schematic for water-based dispersions (based on Fages et al. [17])

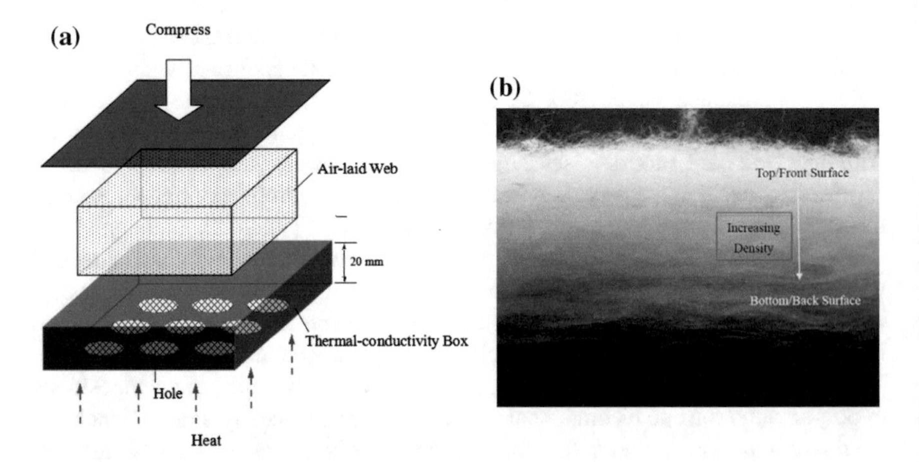

Fig. 6 **a** Setup for compression and single-sided heating of airlaid web. **b** Cross section of the thermal bonded nonwoven fabric (adapted from Zhu et al. [19])

was reflected by it. On the other hand, the loosely packed face allowed sound waves to penetrate the structure and interact with more number of fibres [19].

Meltblowing process of producing nonwoven fabrics involves direct production of fibrous web from polymers using air or another medium that attenuates the fibres. It is usually used where microfibres are required in the nonwoven structure. Meltblown nonwoven fabrics have also been studied and reported for application in sound insulating materials. A layered structure for automobiles that has a thin laminate, composed of polypropylene microfibres produced by meltblowing technique, over a supporting porous foam is described in a patent [20]. Deposition of meltblown fibrous web of polypropylene and polyethylene terephthalate over bulky nonwoven, produced by carding following needlepunching, to produce layered structures has been reported. These structures had faces of fine microfibres that interact first with the sound waves [21].

4 Polymer Composites

Composites are not usually made for sound barrier material but the acoustic insulation properties of them are significantly important because of composites use in buildings and civil structures. Sound barrier properties of various synthetic and natural fibre-based composites are studied by some researchers. Fibres and nonwoven-based polymeric composites are also among the mostly studied structures after the nonwoven fabrics. Even the nonwovens are used as reinforcement with various thermoplastic and thermosetting matrices. Nonwovens are usually used in automobiles flooring, lining materials in various machine structures, etc. whereas polymer composites are significantly used in construction field that demands thermal and sound insulation. The compact surface of composite helps reflecting the sound waves. Furthermore, if the composites are used in conjunction with other sound barrier fillers, then both the sound transmission loss and sound absorption can be enhanced.

Many studies on composites made of textile materials for the purpose of acoustic application follow the path of compression moulding technique. It is a confined moulding process that uses metallic moulds to fabricate the composite sample. Usually the base mould is stationary while the top mould moves towards it and is attached with a hydraulic press to apply pressure. There is a provision to heat both the moulds. Due to the applied heat and high pressure, the prepreg takes the shape of the mould. In textile materials, the thermoplastic components melt or partially melt to adhere to the other components in the prepreg system. Compression moulding technique is used for the materials that have thermoplastic components in it. Other techniques that are less reported in contrast to the acoustic insulation purpose is the vacuum assisted resin transfer moulding (VARTM) where less viscous thermoset resins mixed with hardner and other fillers is allowed to penetrate the samples under vacuum conditions. The presence of vacuum minimizes the amount of void formation in the processed samples.

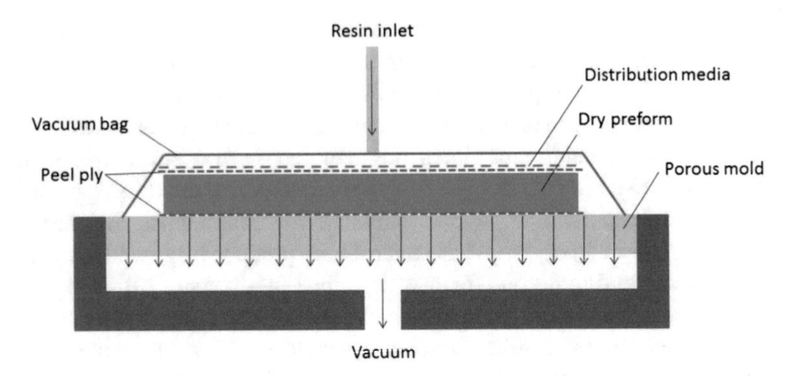

Fig. 7 Schematic of VARTM process using a porous mould (adapted from Yokozeki et al. [23])

Naturally occurring cellulose-based nonwoven, also called bark cloth, obtained from *Ficus natalensis*, *Ficus branchypoda* and *Antiaris toxicaria* trees were investigated for sound absorption properties. Both the individual nonwovens and their layered structures from all the above three varieties were studied for sound absorption properties. Later the layered structures were used to make composite based on epoxy resin using VARTM process. It was observed that the composites made of layered bark cloth lead to reduction in sound absorption coefficient due to reduction in porosity and thickness as VARTM process allows epoxy resin to reach every part of nonwoven reinforcement resulting in compact structure with reduced thickness [22]. Experimental setup of VARTM process using porous mould is shown in Fig. 7 where 10 mm thick porous aluminium plate is used as porous mould.

Layered nonwoven polyester fabric with different number of layers were produced and layers of polypropylene nonwoven were inserted between them to produce thick structures. These were needlepunched and bonded with one layer of thermoplastic urethane. This structure was subjected to compression moulding at 170 °C for 10 min. It was observed that increasing the number of layers resulted in higher sound absorption coefficient, especially at medium and higher frequencies [24]. Composites of glass fibres, polyester, polypropylene and polyvinyl alcohol with polyacrylate were prepared by solution casting. The slurry contained 2 % of fibres and samples with different thickness. Transmission loss characteristics of these composites matched closely to the loss modulus of the structures [25].

In order to improve the sound absorption of composites, the matrix can be made up of porous sound absorbing material. Porous materials perform better in high frequency region. Different amount of blowing agents can be used to attain different pore sizes. Glass fibres when used with porous polyurethane as matrix performed better in sound absorption at medium to high frequency range [26].

Butyl rubber, a viscoelastic material, can be inserted into the composite film to enhance the sound wave damping properties. The secondary vulcanization of the rubber can be carried out with the curing cycle in case of epoxy-based composites. Such structures are called embedded and co-cured composites [27]. Such composites having an embedded viscoelastic character improves sound insulation at those frequencies, also which escape easily through the composite structures and are important in acoustic stealth technology. In the case of lignocellulose-based composites, the viscoelastic properties can be controlled by introduction of polyethylene glycol and glycerol. These hydrophilic molecules enter the fibre structure and replace water molecules that engages in hydrogen bonding [6]. Thus the damping properties, of fibres as well as composites made of them, can be controlled.

Another interesting approach to improve the acoustic properties of composites is to use the voids. Voids are considered as a disadvantage in manufacturing of composites as they are regarded as a flaw and due to high stress concentrations, they may fail the composite easily. However, 3D composites are not very much affected by voids [28]. If the voids do not affect the desired mechanical properties, then they can be introduced into the structure to improve its sound barrier properties. Perforated composite structures are used in combination with other conventional fillers such as vermiculite and perlite, etc. Layered coir fibre mats placed inside the perforated polyester resin composite showed better absorption of sound waves at medium to high frequency region [29].

5 Non-conventional Textile Structures

5.1 Recycled Materials

Waste management and sustained growth are the key areas of interest these days. Enormous amount of polymeric and non-polymeric waste is produced everyday. There are many materials that need separation and are very difficult to recycle due to impurities. Tyre industry uses synthetic and metallic wires with rubber matrix. Textile waste is also not readily recyclable. Thermosetting plastics and different types of foams are also a concern in waste management. One way is to reduce the usage of such materials and other is to reuse these waste materials if recycling is not possible.

Polymeric materials obtained from waste may or may not have desired mechanical properties. Reclaimed fibres, rubber particles, etc. are among some materials that have been used by researchers to make sound barrier products. Waste wool fibres and recycled polyester fibres were used to form nonwovens using needlepunching technique. Both thermal and sound barrier properties were tested for the single- and double-layered fabrics. The woollen nonwovens showed up to 70 % acoustic absorption [30]. Nitrile butadiene rubber obtained from discarded rollers in textile industry was used as matrix for seven-hole polyester fibres. Composites made by compression moulding were tested with varying thickness and

Fig. 8 Tyre fluss with residual rubber (adapted from Maderuelo-Sanz et al. [36])

fibre content for mechanical, dynamic mechanical and sound absorption properties. Sound absorption properties were prominent in composites with 25 % fibres in high frequency region [31]. Cotton and jute waste were used with abandoned polypropylene and thermoplastic polyurethane. Composites were made by mixing in double roller mixer followed by hot press moulding. Experiment claimed that the transmission loss can be achieved with lower thickness if these composites were used [32]. Composites having biopolymer as matrix and cotton as reinforcement, foamed biopolymer with TiO_2 as filler, etc. have also been explored for possible sound barrier properties [33, 34]. All polyester sound absorbing materials obtained from recycled PET fibres and fabrics were used to make dash silencer for automotive and performed similar to conventional polyurethane foam material. The study aimed towards providing the PET recycled material as alternative for non-recyclable polyurethane foam [35]. Fluff obtained from shedding of discarded heavy vehicle tyres was collected that contained fibres of 20–30 μm diameter. This fluff was tested as an entangled mass without imparting resin as well as with resin. Figure 8 includes the microscopic image of the tyre fluff.

Samples showed reduction in sound absorption that contained resin. Though significant absorption was observed at higher frequency in all the samples [36]. An attempt towards exploring agriculture waste for sound absorption was carried out by Sezgin and Küçük [37]. Waste from tea leaf fibres was used as such in bulk mass with and without air cavity. Results show that tea leaf waste provide better sound absorption than polypropylene and polyester-based nonwoven fabrics.

5.2 Shear Thickening Fluids

Shear thickening fluids (STFs) are sticky liquids that become hard under impact loading. This STFs absorb enormous amount of energy in this process. This is the reason for their use in soft body armours. A recent study on exploring the role of

STF applied knitted fabrics in sound insulation has been carried out. Shear thickening fluid based on 30 % silica and polyethylene glycol was used in the study. STF covered the gaps between the yarns and increased the surface density of the knitted fabrics. STF treated knitted fabrics showed almost 10 times loss in transmission as compared to untreated ones [38]. The STF reported in the research was prepared by stirring the mixture of nano silica and polyethylene glycol (PEG). The dispersion underwent through a planetary ball mill for 1–2 h. Stable STF was obtained after subjecting it to vacuum for 2 h at 20 °C. As the STF obtained by this process is highly viscous and cannot penetrate the fabric, a solution of STF in absolute ethanol with equal volume ratio was used to treat the fabric.

In another research, a STF based on tetrapod zinc oxide (ZnO) whisker in PEG was prepared [39]. Tetrapod ZnO was the dispersed phase and PEG was the dispersion medium. These were used in combination with another STF based on nano silica and PEG and applied on single- and double-layered glass fabric after dilution with absolute ethanol. It was reported in the study that the treated fabrics displayed better sound transmission loss due to increase in tightness of fabrics and increased areal density.

5.3 Resonance Membrane

Porous sound absorbing materials work efficiently at higher frequency but their performance at low frequency is very poor. An invention aims towards this particular problem and suggests a layer of resonance membrane made up of nanofibres. Membrane is capable of dissipating the sound energy by achieving resonance at lower frequency. This membrane is made up of nanofibres that have diameter of 600 nm and surface weight of 0.1–5.0 g/m^2. The membrane can be placed between 2 and 3 layers of normally carded webs. This setup can be used in automotive, aviation and buildings, etc. [15]. Resonance frequency of nanofibrous resonance membrane decreases as the weight per unit area increases. There is an increase in its resonance frequency as the diameter of the nanofibre decreases [40]. Magnified image of an electrospun nanofibrous resonance membrane is shown in Fig. 9.

A detailed method of production of layered card fibrous web structures using these membranes is mentioned in another patent and the production setup is mentioned in Fig. 10 [41].

5.4 Carbon Fibre Felts

In order to restrict low frequency noises, various other approaches such as introduction of air cavity, hybrid structures having textile and nontextile components, etc. are adopted. Activated carbon fibres have larger surface area, interconnected micro-pores and thus an ideal candidate for sound absorbing material. A decrease in

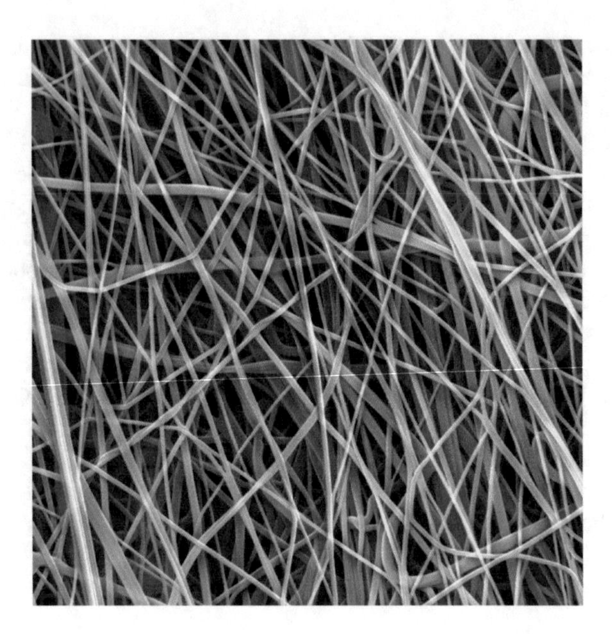

Fig. 9 Microscopic image of resonance membrane formed by electrostatic spinning (adapted from Kalinová [40])

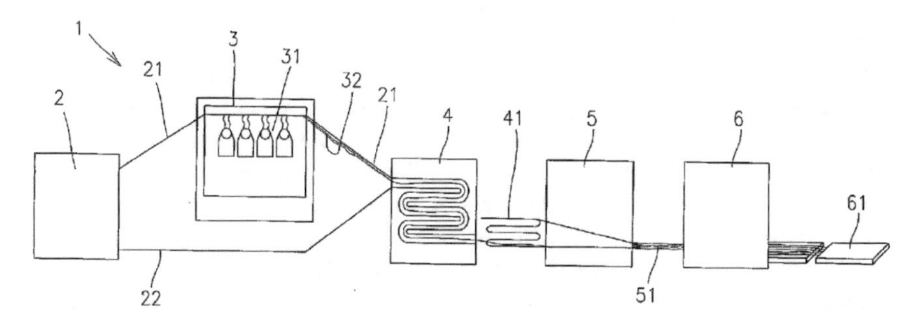

Fig. 10 Schematic diagram of production process of layered fabric embedded with resonance membrane in the middle. The parts are as follows *1* production line, *2* carding machine, *21* upper layer of the card fibrous Web, *22* lower layer of the card fibrous Web, *3* device for production of nanofibres through electro static spinning, *31* spinning compartment, *32* layer of nanofibres, *4* cross laying device, *41* non-reinforced layered sound absorptive fabric, *5* hot-air chamber, *51* layered sound absorptive nonwoven fabric, *6* cutting device, *61* panel (adapted from Stránská et al. [41])

production costs and advanced carbonization and activation of carbon fibres has further pushed its use in less sophisticated applications also. The production process of carbon fibres includes oxidation, carbonization and activation. Oxidation step includes heating of precursor in the presence of air. Carbonization is used to remove non-carbon impurities and is carried out in nitrogen atmosphere. Activation part introduces etching and pore formation on surface. A study reported by Shen and Jiang [5] describes use of carbon fibre felts which were made from viscose nonwoven fabrics produced by needlepunch technique and used for sound absorption. The viscose nonwoven mat was dried and put into 5 % solution of ammonium dihydrogen phosphate for half an hour. Preoxidation was carried out in an oven at 300 °C. Carbonization was carried out by further heating it up to 800 °C in nitrogen atmosphere. Vapour in the impedance oven was used for activation of felts for 8 min followed by natural cooling. The optimized position of carbon fibre felts with respect to air cavity and backing material is discussed in this study. It is seen that the role of air cavity is very significant, especially in the low frequency region. Same research group also discussed the effect of thickness and bulk density of felts along with fibre diameter on sound absorption in another study [42]. The group further carried out a detailed study of process parameters of producing carbon fibre felts and its effect is further studied in relevance to sound absorption properties. It was observed that the increase in carbonization temperature, activation temperature and activation time positively affected the sound absorption [43].

5.5 *Fabric Sleeves*

Textile structures find significant application in interiors and acoustic absorption in automobile sector. Sleeving is applied over the tubular parts in the machinery to minimize the sound while in operation. Textile sleeves are preferred over other materials due to long lasting, textured, soft and bulky nature. Original sleeves consisted of interthreaded textured yarns and were soon replaced by knitted sleeves that can be stretched and installed. As the knitted sleeves suffer from inability to fully regain their shape after application, leading to sag and displacement from original position, alternatives were searched. Braided sleeves also tend to sag under insufficient tension. The monofilaments cross over each other and form a hard spot that is undesirable as it do not contribute to the sound absorption. A process of making braided sleeves without hard spots is mentioned in an invention [44]. This process describes use of DREF 3 produced core-sheath yarns with monofilament fibre in core and staple fibres on sheath for producing braided sleeves. As the yarns produced by this process are resilient, soft and bulky, they give rise to a braided structure with no tight spots on the crossover points. The final structures were subjected to heat set in desired condition to stabilize them.

5.6 *Other Textile Materials*

Carpets are one of the widely used textile materials and some studies are also carried out to analyze the effect of pile height, backing material, density, air permeability, etc. on its sound barrier properties.

Carpets can be needle felts, woven, knotted or tufted. Tufted carpets are usually studied regarding acoustic properties. These carpets have a pile structure that is attached to primary backing which in turn is attached to a heavy construction called secondary backing. Tufted carpets have complex sound absorbing mechanism. The absorbance by porous pile structures that consumes sound energy by internal friction of walls of pores and the resonance absorption that relates to backing material and its resonance frequency that has a certain maxima at some frequency. The pile height, density, material, backing material's density and construction all affect the sound barrier properties of tufted carpets. As the studies carried out in this regard are very less, it is not justified to come to a conclusion as to what parameters influence the sound barrier properties.

A plain knitted fabric cannot provide enough sound barrier properties due to its limited thickness and density. Although other knitted structures such as rib, interlock, etc. are thicker but they also do not provide sufficient sound barrier properties. Thus, use of a spacer fabric can be beneficial. A spacer fabric consists of two layers of knitted structure that are connected by monofilament but with some space between the two layers. Fabric thickness is dependent on length of connecting yarns. Spacer fabrics are produced on warp knitting machines (Fig. 11). Such 3D spacer structures are less flexible and display poor drape properties. At the same time, they are less costly and unaltered by moisture fluctuations [45].

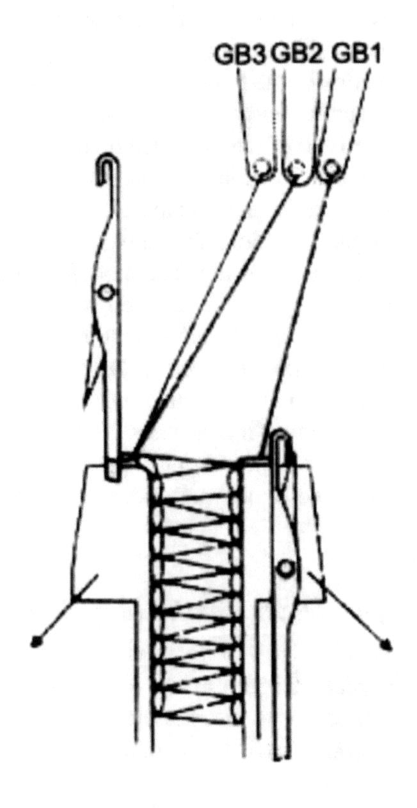

Fig. 11 Knitting setup used for production of spacer fabric (adapted from Zhang et al. [46])

3D textile structures are among the less explored materials for acoustic insulation. The studies on 3D spacer fabrics for acoustic insulation are still in initial phase and therefore the emphasis in all these studies is primarily to understand the role of face and back layer's density, air permeability through the spacer structure and thickness of the spacer fabric. There are no dedicated routes for production of spacer fabrics exclusively for acoustic insulation application. Liu and Hu [47] studied the noise absorption coefficient for weft and warp knitted spacer fabrics and concluded that a combination of both can lead to better results than placing an air cavity behind the structure. Tuck spacer fabrics were studied by Dias et al. [48] to explore its application in reducing noise in automobiles. In a recent research, 3D spacer fabric composites were prepared by hand lay-up technique using epoxy resin. Release agent was applied on inner side of the mould before placing the samples. Both sides were applied with the epoxy resin one by one. Samples were allowed to cure under optimum conditions and tested for transmission loss. It was observed that structures having '8'-shape had higher fabric density and thus, showed higher transmission loss than '88'-shaped fabrics [49]. A 3D spacer fabric has been shown in Fig. 12.

Three-dimensional orthogonal woven fabric is manufactured using multiwarp yarn system. In these structures, the straight warp and weft yarns are bonded together by binder yarns. These binder yarns stablize the structure in the third dimension [49]. Three-dimensional composites from orthogonally woven fabrics are superior to two-dimensional fabric composites due to lesser tendency of delamination and higher in-plane mechanical properties. Usually thermoplastic matrices find it difficult to penetrate the structure of 3D woven fabrics due to their higher viscosity. Keeping this in mind, alternate way of producing composites using 3D orthogonal woven fabrics and thermoplastic matrix was carried out in a study [28]. Three-dimensional orthogonal woven composites made up of PLA wrapped ramie fibres were studied (Fig. 13).

Wrapping up of PLA fibres on ramie was carried out on a fancy twisting machine. Orthogonal 3D weaving was carried out on a computer controlled loom. Whole assembly was heated at 165 °C and pressed for 3 min at 1–2 MPa pressure. Voids were deliberately introduced into the structure because they do not affect the mechanical properties of 3D woven composites and provide a tortuous path for higher sound absorption.

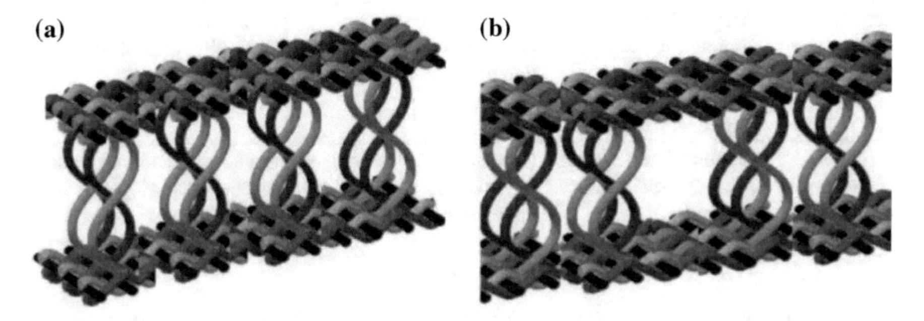

Fig. 12 3D spacer fabric of **a** "8"-shaped, and **b** "88"-shaped pile structure (adapted from Haijian et al. [49])

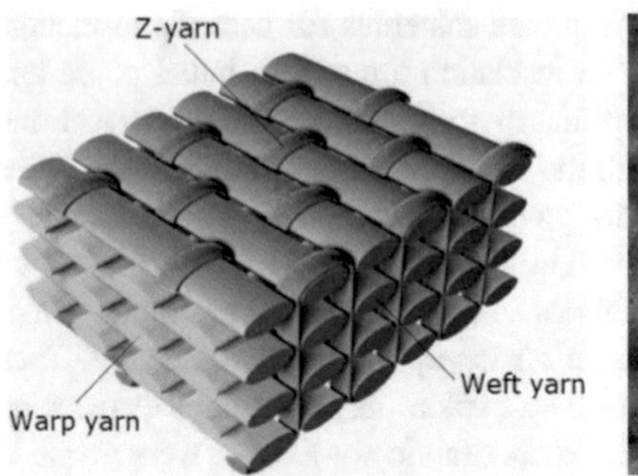

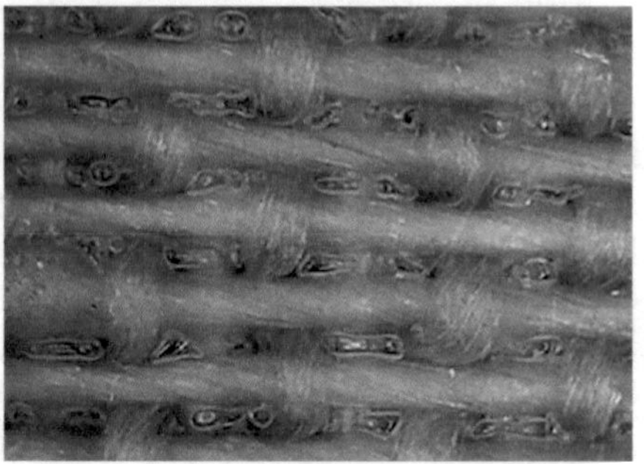

Fig. 13 Schematic of orthogonal 3D woven fabric and the surface of composite made by PLA and ramie fibres (adapted from Zhou et al. [28])

Table 1 Brief comparison between various processes used to produce textile materials for acoustic application

Production process	Features	Advantages	Disadvantages
Weaving	Produces thin structures, Slow production speeds	Simplest of all methods, cheaper than other processes	Lesser window to optimize the factors affecting sound insulation, limitation on fabric thickness
Nonwoven	Porous textile structures are produced, fast production	Extent and size of porosity can be easily controlled, different fibres can be processed simultaneously, most widely studied technique	Limitation on thickness of fabrics, only planar structures can be prepared
Polymer composites	Rigid structures, reflects most sound waves	Textile waste can be utilized, vast variety of combinations can be produced, desired shape can be fabricated, second most widely studied technique	Reduces porosity of product, flexibility is lost
Knitting	Warp and weft knitting produce open structures, spacer knitting is less common	Versatile in process parameters, simple non-planar structures are possible	Spacer knitting reduces flexibility, only spacer construction is dimensionally stable
Hybrid structures	Combination of two or more structures or approaches	Effective in high as well as low frequency range, highly versatile for end application	Sound insulation depends on sequence of placement of components

A brief comparison of various processes used for the fabrication of acoustic textiles is given in Table 1.

6 Conclusion

Textile materials are flexible, less costly and easy to customize to meet the desired end use. Due to these advantages of textile materials, they are explored for sound insulation as well as sound absorption properties. Though textiles are not a very good candidate for sound barrier materials, their inherent porous structure gives some advantage in sound absorption. Natural hollow fibres and fine synthetic fibres with specially shaped cross section are usually explored for making nonwoven fabrics and polymer composites. Most of the textile materials used are nonwoven fabrics. Though spunbonding, electrospinning and wet lay are also reported for nonwoven production for acoustic insulation purpose, needlepunching technique is the most popular. Needlepunch density, speed of web, web orientation, etc. are the parameters that are widely analyzed for their effect on sound barrier properties.

Polymer composites are not usually manufactured to provide acoustic barrier properties. The use of composites in buildings, workplaces, etc. demands better sound and thermal insulation. Therefore, various studies to optimize the fibre type, fibre content, void content, etc. for producing composites are studied. Recycled materials and non-conventional activated carbon felts are also explored to provide better acoustic properties.

Manufacturing methods for all the above products are similar to the conventional ones as the research in textiles for acoustic insulation mainly targets optimization of different process variables rather than modifying the existing production setup. Nonwoven and polymer composites have been found to be the most frequently used techniques for production of sound barrier materials made of textiles. All other methods can be further explored by researchers towards creation of acoustic textiles.

References

1. Zhang X, Liu C (2013) Study of a sound absorbing polyurethane based on porous composite material. Appl Mech Mat 275–277:1623–1627
2. Lueg P (1936) Process of silencing sound oscillations. U.S. Patent. 2,043,416, 9 Jun. 1936
3. Soltani P, Zarrebini M (2013) Acoustic performance of woven fabrics in relation to structural parameters and air permeability. J Text Inst 104:1011–1016
4. Suvari F, Ulcay Y, Maze B, Pourdeyhimi B (2013) Acoustical absorptive properties of spunbonded nonwovens made from islands-in-the-sea bicomponent filaments. J Text Inst 104:438–445
5. Shen Y, Jiang G (2014) Sound absorption properties of composite structure with activated carbon fiber felts. J Text Inst 105:1100–1107
6. Le Guen M, Newman R, Fernyhough A, Staiger M (2014) Tailoring the vibration damping behaviour of flax fibre-reinforced epoxy composite laminates via polyol additions. Compos Part A Appl Sci Manuf 67:37–43
7. Liu X, Li L, Yan X, Zhang H (2013) Sound-absorbing properties of Kapok fiber nonwoven composite at low-frequency. Adv Mat Res 821–822:329–332

8. Hassanzadeh S, Hasani H, Zarrebini M (2014) Analysis and prediction of the noise reduction coefficient of lightly-needled Estabragh/polypropylene nonwovens using simplex lattice design. J Text Inst 105:256–263
9. Na Y, Lancaster J, Casali J, Cho G (2007) Sound absorption coefficients of micro-fiber fabrics by reverberation room method. Text Res J 77:330–335
10. Voronina N (1994) Acoustic properties of fibrous materials. Appl Acoust 42:165–174
11. Tascan M, Vaughn EA (2008) Effects of fiber denier, fiber cross-sectional shape and fabric density on acoustical behavior of vertically lapped nonwoven fabrics. J Engg Fibers Fabr 3(2):32–38
12. Tascan M, Vaughn E (2008) Effects of total surface area and fabric density on the acoustical behavior of needlepunched nonwoven fabrics. Text Res J 78:289–296
13. Harting H (2005) Fibre for an acoustic insulating material, especially for sound dampers compressed air devices. U.S. Patent Application 11/658,291
14. Nayak R, Padhye R, Kyratzis IL, Truong YB, Arnold L (2012) Recent advances in nanofibre fabrication techniques. Text Res J 82:129–147
15. Kalinova K, Sanetrnik F, Jirsak O, Mares L (2006) Layered sound absorptive non-woven fabric. U.S. Patent Application 11/911,135
16. Soltani P, Zerrebini M (2012) The analysis of acoustical characteristics and sound absorption coefficient of woven fabrics. Text Res J 82:875–882
17. Fages E, Cano M, Girones S et al (2012) The use of wet-laid techniques to obtain flax nonwovens with different thermoplastic binding fibers for technical insulation applications. Text Res J 83:426–437
18. Berto L, Romina DR, Jesús A, Vicent S (2012) Reuse of textile powder remainders for acoustic applications using the Wet-Laid technology. In Acoustics 2012, April 2012
19. Zhu W, Nandikolla V, George B (2015) Effect of bulk density on the acoustic performance of thermally bonded nonwovens. J Engg Fabr Fibers 10(3):39–45
20. Vanbemmel WR, Anthony CV, Thorsten A (2012) Sound absorbent thin-layer laminate. U.S. Patent No. 6,720,068, 13 Apr. 2004
21. Lee B, Ko J, Han S (2010) Characteristics of PP/PET bicomponent melt blown nonwovens as sound absorbing material. Adv Mater Res 123–125:935–938
22. Rwawiire S, Tomkova B, Gliscinska E et al (2015) Investigation of sound absorption properties of bark cloth nonwoven fabric and composites. Autex Res J
23. Yokozeki T, Yamazaki W, Kobayashi Y (2015) Investigation into property control of VaRTM composites by resin infusion process. Adv Compos Mater 24(6):495–507
24. Huang C, Lin J, Chuang Y (2013) Manufacturing process and property evaluation of sound-absorbing and thermal-insulating polyester fiber/polypropylene/thermoplastic polyurethane composite board. J Indus Text 43:627–640
25. Chen R, Yao M, Yang P, Wang X (2014) Investigation of the damping properties and the sound insulation performances of fiber-reinforced composites. Ferroelectrics 470:194–200
26. Zhang X, Liu C (2013) Study of a sound absorbing polyurethane based on porous composite material. Appl Mech Mater 275–277:1623–1627
27. Liang S, Xiu Y, Wang H (2012) A research on sound insulation characteristics and processing of the embedded and co-cured composite damping structures. J Compos Mater 47:1169–1177
28. Zhou N, Geng X, Ye M (2014) Mechanical and sound adsorption properties of cellular poly (lactic acid) matrix composites reinforced with 3D ramie fabrics woven with co-wrapped yarns. Indus Crops Produc 56:1–8
29. Zulkifli R, Nor M, Tahir M et al (2008) Acoustic properties of multi-layer coir fibres sound absorption panel. J Appl Sci 8:3709–3714
30. Patnaik A, Mvubu M, Muniyasamy S et al (2015) Thermal and sound insulation materials from waste wool and recycled polyester fibers and their biodegradation studies. Energy Build 92:161–169
31. Zhou X, Jang S, Yan X et al (2013) Damping acoustic properties of reclaimed rubber/seven-hole hollow polyester fibers composite materials. J Compos Mater 48:3719–3726

32. Zuo M, Lv L, Wei C et al (2013) Sound insulation properties of abandoned fibers composites. Adv Mater Res 821–822:1189–1192
33. Hassan N, Rus A (2013) Influences of thickness and fabric for sound absorption of biopolymer composite. Appl Mech Mater 393:102–107
34. Rus A, Normunira N, Rahim R (2014) Influence of multilayer textile biopolymer foam doped with titanium dioxide for sound absorption materials. Key Engg Mater 594–595:750–754
35. Watanabe K (1999) Development of high-performance all-polyester sound-absorbing materials. Japan Soc Automot Engg Rev 20:357–362
36. Maderuelo-Sanz R, Nadal-Gisbert A, Crespo-Amorós J, Parres-García F (2012) A novel sound absorber with recycled fibers coming from end of life tires (ELTs). Appl Acoust 73:402–408
37. Ersoy S, Küçük H (2009) Investigation of industrial tea-leaf-fibre waste material for its sound absorption properties. Appl Acoust 70:215–220
38. Li S, Wang Y, Ding J et al (2014) Effect of shear thickening fluid on the sound insulation properties of textiles. Text Res J 84:897–902
39. Wang Y, Zhu Y, Fu X, Fu Y (2014) Effect of TW-ZnO/SiO2-compounded shear thickening fluid on the sound insulation property of glass fiber fabric. Text Res J 85:980–986
40. Kalinová K (2011) Nanofibrous resonant membrane for acoustic applications. J Nanomater 2011:1–6
41. Stránská D, Mares L, Jirsák O, Kalinová K, Stranska D, Jirsak O, Kalinova K (2008) Production method of layered sound absorptive non-woven fabric. U.S. Patent Application 12/522,410
42. Shen Y, Jiang G (2013) Effects of different parameters on acoustic properties of activated carbon fiber felts. J Text Inst 105:392–397
43. Shen Y, Jiang G (2015) The influence of production parameters on sound absorption of activated carbon fiber felts. J Text Inst 1–6. doi:10.1080/00405000.2015.1097083
44. Flasher GL (1999) Sound absorbent fabric sleeves. U.S. Patent 5,866,216, 2 Feb 1999
45. Monaragala RM (2011) Knitted structures for sound absorption. In: Au KF (ed) Advances in knitting technology, 1st edn. Elsevier, pp 262–28545
46. Zhang LZ et al (2012) Three-dimensional computer simulation of warp knitted spacer fabric. Fibr Text East Eur 56–60
47. Liu Y, Hong H (2010) Sound absorption behavior of knitted spacer fabrics. Text Res J 80:1949–1957
48. Dias T, Monaragala R, Needham P, Lay E (2007) Analysis of sound absorption of tuck spacer fabrics to reduce automotive noise. Meas Sci Tech 18:2657–2666
49. Haijian C, Kejing Y, Kim Q (2013) Sound insulation property of three dimensional spacer fabric composites. Fibr Text East Eur 38(3):304–308
50. Luo Y, Lv L, Sun B et al (2007) Transverse impact behavior and energy absorption of three-dimensional orthogonal hybrid woven composites. Compos Struct 81:202–209

Acoustical Test Methods for Nonwoven Fabrics

Mevlüt Taşcan

Abstract Acoustical properties of nonwoven fabrics are measured using several different test methods. These important test methods measure physical, mechanical and acoustic properties, which are important for the evaluation, characterization and differentiation of nonwoven fabrics for acoustical applications. Even the basic physical properties of nonwovens play a very important role in acoustics. Physical, mechanical and acoustical properties of nonwoven fabrics are measured using several standards that are defined in ASTM, ISO, ANSI and SAE test standards. Some other measurement techniques that are not included in any standard test methods can also be found in literature. In this chapter, important standard or nonstandard test methods related to acoustical properties of nonwoven fabrics are explained and described. Important physical properties of nonwoven fabrics for acoustical applications are areal density (mass), volumetric density, porosity, particle size distribution, tortuosity and thickness. Density parameters are the most important parameters for the differentiation of the nonwoven fabrics. Porosity also plays a major role in acoustical properties because each pore in the fabric becomes an absorptive material for the sound waves during sound wave–nonwoven fabric interaction. Mechanical parameters that are related to acoustical properties of nonwoven fabric are tensile properties and Young's modulus. Acoustical absorption properties of nonwoven fabrics are determined using sound absorption coefficient, sound transmission loss, noise reduction coefficient, sound impedance and airflow resistivity parameters. These parameters could be measured using many different test methods. These methods are explained and related ASTM, ISO, ANSI and SAE test standards are discussed in this chapter. Especially impedance tubes and reverberation rooms are used for most of the acoustical test methods discussed in this chapter.

Keywords Acoustical test methods · Nonwoven fabrics · Acoustical absorption · Porosity · ASTM standards · ISO standards

M. Taşcan (✉)
Zirve University, Gaziantep, Turkey
e-mail: mevlut.tascan@zirve.edu.tr

R. Padhye and R. Nayak (eds.), *Acoustic Textiles*, Textile Science and Clothing Technology, DOI 10.1007/978-981-10-1476-5_6

1 Evaluation Methods for Acoustic Materials

1.1 Introduction

Acoustic science is mostly related to the physical properties of the fibres. Since the acoustical waves are pressure waves, acoustical absorbing or insulating materials absorb and/or insulate the acoustical waves with the energy transformation. Most of the time, when the acoustical waves meet with the acoustical absorbing or insulating textiles, the energy of the acoustical waves transforms into heat and kinetic energy. Therefore, this energy change may be investigated at the macroscopic scale rather than the atomic scale. So when the acoustical test methods of nonwoven fabrics are researched, mostly physical test methods are used.

Physical, mechanical and acoustic properties are important for the evaluation, characterization and differentiation of nonwoven fabrics for acoustical applications. Even the basic physical properties of nonwovens play very important roles in acoustics.

In this chapter, first the physical, mechanical and acoustical absorption properties of nonwoven fabrics and the test methods used for their evaluation will be explained. Each parameter is selected based on its importance and effect in acoustical absorption and insulation. Some other parameters were selected mostly because of their being an indicative parameter for nonwoven fabrics. The physical properties selected were areal density, thickness, volumetric density, porosity and tortuosity; mechanical properties selected were tensile properties and Young's modulus; acoustical absorption properties selected were absorption coefficient, sound transmission loss, noise reduction coefficient, sound impedance and airflow resistivity.

1.2 Physical Properties

Important physical properties of nonwoven fabrics for acoustical applications are areal density (mass), volumetric density, porosity, particle size distribution, tortuosity and thickness. Density parameters are the most important parameters for the differentiation of the nonwoven fabrics. Most of the time, the areal density, in weight per unit area, is used for naming the nonwoven fabrics, whereas higher weight fabrics have the higher areal density. For defining the nonwoven fabrics, most of the time thickness does not play much of a role. Thickness is very important for this chapter because most of the researchers in acoustical textile area accept the thickness parameter of the nonwoven fabrics as the most important parameter for thermal and acoustical insulation and absorption. Porosity is also very important for the acoustical absorption and insulation of nonwoven fabrics because more porous nonwoven fabrics have generally lower weight but at the same time, allow the acoustical waves to enter the nonwoven fabrics. When the acoustical wave enters the nonwoven fabric, waves interact with the fibres and each fibre behaves as different absorbing and insulating material.

1.2.1 Areal Density

Areal density is also known as the mass of the textile material. Areal density could be defined as the weight of the certain area of the fabric. In general, textile materials are also called two-dimensional materials because they have very small thickness. Especially knitted, woven, spunbond and meltblown nonwoven fabric could be included in this category. For most of the acoustical applications (acoustical absorption), textile-based nonwoven fabrics are used, and these fabrics have relatively high thicknesses. But still, areal density parameter is very important because these fabrics are made as rolls, and the production speed is defined as square metre per unit time. Therefore, areal density (mass) of the textile material is a very distinctive physical property for all acoustical textiles.

Areal density could be easily measured using balance with a minimum 0.01 g sensitivity. At least a 10 cm × 10 cm (0.1 m × 0.1 m) area should be cut from the roll, and the weight of this fabric should be measured in grams. Textile materials are cut using dies with certain area. At least five different pieces of fabric should be cut from different parts of the roll. Then the weight is divided by the area of the fabric and the final value is found as g/m^2. The final average of the five results is calculated. The standard test method used to find the areal density of the textile material is ASTM D3776/D3776M, and it is known as "Standard Test Methods for Mass Per Unit Area (Weight) of Fabric." Based on region, the unit of areal density could be grams per square metre (g/m^2) or ounces per square yard (oz/yd^2).

1.2.2 Thickness

The thickness of the textile materials is generally very small for knitted and woven fabrics. The thickness of the nonwoven fabrics made from staple fibres is relatively much higher, and it could increase up to 10 cm for the insulation applications.

Thickness is a very distinctive parameter for acoustical insulation and absorption of textiles. In 1980s and 1990s, many of the researchers working on acoustical absorption used the thickness of the textile as an indicative factor for acoustic absorption. Thickness and sound absorption were accepted as directly related to each other.

As known, most of the textile-based sound absorption materials are soft and relatively thick materials. Because of their softness, these materials do not have perfect structural integrity. So the thickness measurements of textiles are not as easy as the thickness measurements of solid materials. Thickness is measured using ASTM D1777 for textiles and ASTM D5729 and ASTM D5736 for nonwoven fabrics. The methods are basically measuring the thickness using some weight. As mentioned in these standards, the thickness gauge (presser foot) should push the fabric with 0.21 kPa pressure so the related weight and the presser dimensions could be selected using this pressure. Because the textile-based acoustic materials are not very uniform, at least 10 measurements should be done and the average is taken as reliable data.

1.2.3 Volumetric Density

Volumetric (bulk) density is a very distinctive parameter for all types of materials, including nonwovens. The unit of the volumetric density is grams per centimetre cube (g/cm^3). Textiles used for acoustical applications are solids, and volumetric density is found measuring the weight and the three dimensions of the fabric. Because the textile-based acoustic materials are not very uniform, at least 10 measurements should be done and the average is taken as reliable data.

1.2.4 Porosity

Pore is defined as the voids (air spaces) in the textile materials. As mentioned before, sound absorbers should have air gaps in the structure so the sound energy could change into thermal and kinetic energy. So porosity plays a very important role for sound absorption.

In general, there are two types of measurements of pore structures of the textiles: one is the total reachable porosity of the textiles and the other is the pore size distribution of the textiles. Total reachable porosity could be found using the density ratio of the bulk textile material to the polymer produced from the textile material as in the following equation:

$$\text{Porosity} = \frac{(1 - \text{Density of the fabric})}{(\text{Density of the fibre})}$$

The second parameter used for porosity is the pore size distribution. This parameter is also very important for the acoustic properties of the material because when the pore size is smaller, there will be more numbers of fibres in the specific area, so the acoustic absorption will be higher. So higher pore sizes at the same volumetric density allow less sound absorption by the material. Pore sizes of the material could be found using extrusion flow porometry, extrusion porosimetry, mercury intrusion porosimetry, non-mercury intrusion porosimetry and gas absorption techniques. One of the measurement devices used to find pore structure of the material is porosity measurement instrument (PMI). Average bubble point diameter and bubble point diameter distributions directly show the pore structure of the fabric. The working principle is that material is dipped in the liquid with very low surface energy so the air inside the material goes out as the applied pressure is increased slowly. So the first air comes out from the big pores. While the pressure is slowly increased, air from the smaller pores goes out of the material and the volume of each air amount results from pore sizes in the material.

In the case of composites prepared from natural materials such as hemp shives and agricultural waste, the intraparticle porosity and the inter-porosity give the total porosity of the material. The intraparticle porosity can be calculated by using optical image and 3D tomography. The total porosity can be measured by a porosimeter [1].

The air permeability of porous materials can be measured for comparison of relative porosity of various materials. Air permeability is a measure of how well the fabric allows airflow through it under a differential pressure between the two surfaces (face and back). Air permeability is defined as the volume of air in millilitres, which is passed in one second through 1 cm^2 of a fabric at a pressure difference of 10 mm head of water. During the test, the specimen is clamped over an air inlet of the test apparatus and air is sucked through it by means of a pump. The air valve is then adjusted to give a pressure drop across the fabric of 10 mm head of water and the airflow is then measured using a flow meter. Five to ten specimens can be used for each material and the mean airflow in cubic centimetre per square centimetre per second is reported. ISO 9237: 1995 (Determination of the permeability of fabrics to air) test standard can be used for the same.

The porosity values play a major role in the air permeability of fabrics. The porosity of various structures can be measured as a volume fraction of the void (empty) place in the structure. It is expressed as a fraction of the volume of empty place over the total volume, which can range between 0 and 1. In the case of a fibrous structure such as fabric, the empty space is indicated by the space not covered by the yarns (or fibres in a nonwoven). These places will facilitate the passage of air through the structure. The porosity of fabric samples can be measured from the optical (microscopic) images by using suitable software such as Image J. The software calculated the pore volume fraction present in the structure by image analysis. Average of 10 readings from three to four different images can be taken as the porosity of the fabric samples.

1.2.5 Particle Size Distribution

It is essential to measure the particle size and size distribution while preparing the composites of natural materials such as hemp shives, agricultural waste, wood-based material and rice straw with synthetics. The size and size distribution can be measured and calculated by optical microscope or scanning electron microscope images and analyzing the images with appropriate software. As the sizes of the constituent material vary a lot, sufficiently high reading should be taken to represent all the size ranges.

1.2.6 Tortuosity

Tortuosity is used to quantify the inertial coupling between the fluid and solid phases of a porous material. In other words, tortuosity is the length of the liquid or

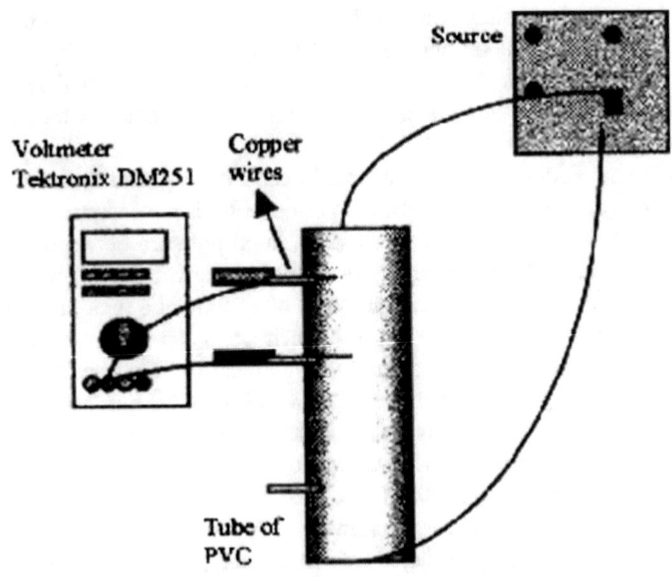

Fig. 1 Experimental setup for tortuosity measurement used by Gerges and Balvedi [2]

gas going through the textile material. It is measured by the method based on electric conductivity, as shown in Fig. 1. The sample of porous material is mounted inside the tube, completely filled with a conductive liquid. The tube is closed at either end by electrodes, and an electric field is generated. Gerges and Balvedi [1] defined tortuosity as given in Eq. (1).

$$q = \varepsilon \frac{\sigma_s}{\sigma_f} \tag{1}$$

where q is tortuosity (cm/cm), ε is the porosity (cm^3/cm^3) of porous material, σ_s is the electric resistivity of porous material ($\Omega \cdot m$), and σ_f is the electric resistivity of fluid ($\Omega \cdot m$) [2].

Tortuosity can also be defined as the total distance of the air travels to pass the material divided by the thickness of the material. Therefore tortuosity is a unitless (distance/distance) parameter.

1.2.7 Compression

The compression properties of textile structure can affect the sound absorption. The compression of fibrous structure results in decrease in thickness and porosity. Furthermore, tortuosity and airflow resistivity values are increased. The major reason for reduced sound absorption by the textile material can be attributed to reduction in the thickness. The compression plays a major role in the case of acoustic textiles in car interior. Due to the weight of the passengers, the compressed thickness of materials such as seat cover and carpet can lead to reduced sound absorption.

The universal tensile tester "Instron" can be used for compression test to determine the behaviour of materials under compressive loads. The textile materials are compressed between two sets of parallel planes and the compressional thickness is monitored at various loads. Other parameters such as elastic limit, yield point, compressive strength and yield strength can be calculated from compressive stress and strain curves of the materials.

Standards such as ASTM D 6571-01 (Standard Test Method for Determination of Compression Resistance and Recovery Properties of highloft Nonwoven Fabric Using Static Force Loading) can be used for compression testing. This standard describes the measurement of compressive resistance and recovery properties of any highloft nonwoven fabric using a simplistic and economical applied static weight loading technique.

Fabric compression can also be measured by Kawabata Evaluation System (KES) using compression module [1, 2]. KES measures the degree of compressional force a fabric can sustain in certain direction before the fabric buckling occurs. Parameters such as linearity of compression thickness curve (LC), compressional energy (WC) and compressional resilience (RC) are measured by KES [3].

1.3 Mechanical Properties

Mechanical properties of nonwoven fabrics include, but are not limited to, strength, force at break, elongation at break, Young's modulus, bursting strength and tearing strength. All of these parameters could be measured using a universal testing machine (UTM) such as Instron.

Mechanical properties of acoustic textiles are important because these specific properties of the materials determine if the textile could be used in certain applications or not. Acoustic textiles are used mostly in automotive and building industries to absorb and/or insulate unwanted sound, which is called noise. So these materials need substantial mechanical properties.

1.3.1 Tensile Properties

Tensile strength is defined as the maximum tensile stress a material can withstand before it fails. The measurement of tensile strength is done using UTM. In general, the measurement procedure is as follows: fabric is drawn from its two directions: machine direction and cross-machine direction in the specific dimension as per the standard and tested. Because the fibre orientation is not distributed randomly on some of the nonwoven fabrics, fabrics are anisotropic materials. Fabric also could be measured between the directions (45° on both machine and cross-machine directions).

The grab tensile tests are done according to the ASTM D5034 (grab method). In this method, fabric is drawn from its two sides and the force at the break point is named as the strength of the nonwoven fabric. In the grab tensile strength measurement method, fabric is cut 5 cm by 10 cm and the fabric is clamped from two sides at 2.5 cm in the middle. So when the force is applied during the test, the jaws draw the fabric and the fabric on the sides starts to move to the middle. At the time of the break, the total elongation of fabric is divided to the first length of the fabric (10 cm), and the result is taken as the percent elongation at the break of the fabric.

Another tensile strength measurement in ASTM standards is the strip method. The number of the test standard is ASTM D5035. In this method, the jaws of the UTM hold the whole width of the fabric since the fabric test dimensions are 2.5 × 10 cm.

The tearing strength of the nonwoven fabric is measured using ASTM D1424 and ASTM D2261 standards. The Elmendorf tearing test device (falling pendulum) is used for the ASTM D1424 standard. UTM is used for the ASTM D2261 method and the name of the method is the "tongue tear test." The test specimen is cut into 5 cm by 10 cm pieces and the specimen is cut from the middle of the width (5 cm dimension) by 5 cm. Then one side of the fabric is locked to one jaw and the other side is locked by the other jaw. Then the test is started, the nonwoven fabric specimen is torn by the device at a constant rate of elongation, and the force is measured. In the Elmendorf test, the specimen is prepared using the same dimensions the same way. Then the pendulum at the standard weight falls and the force is measured as the result of the test.

Another type of strength property of nonwovens is bursting strength. This bursting strength is defined as the amount of pressure that nonwovens can hold before they burst. Bursting strength is measured using ASTM D3786 (diaphragm bursting test) and ASTM D3787 (ball bursting test). Both tests measure the same value. The diaphragm bursting test uses a rubber diaphragm, and liquid oil bursts the fabric using the stress applied by the liquid oil. The ball bursting test uses a steel ball that has a 2.5 cm diameter, and this ball pushes the fabric until the fabric bursts.

Creep is another important parameter for nonwoven fabrics especially for geotextile applications. Creep is a type of deformation. It is the tendency of a nonwoven fabric to slowly move or deform permanently under the influence of stresses. It is measured using a UTM machine. In general, the defined amount of stress is applied to the fabric, and then the fabric starts to come back to its original position with time. So the applied stress and the time are the distinctive parameters for the creep behaviour of nonwoven fabric. Creep is measured using ASTM D5262 and D6992 standards.

1.3.2 Young's Modulus

Young's modulus is defined as the ratio of linear stress to linear strain. It is also measured using a UTM, and Young's modulus is calculated by dividing stress to strain in the elastic region. As known from the literature [4–6], acoustical absorption transforms the acoustical energy into thermal energy and kinetic energy. Kinetic energy on acoustical absorption of nonwoven fabrics is defined as the movement or vibration of fibres. Vibration of fibres directly related to the Young's modulus of these fibres. Therefore, Young's modulus is a very important parameter for acoustical absorption of nonwoven fabrics.

1.4 Acoustical Absorption Properties

Acoustic absorption is the amount of sound energy loss when the sound energy interacts with nonwovens. Nonwoven fabrics are used to absorb and eliminate the unwanted sound energy for any application. Acoustic absorption properties of nonwovens could be measured in several ways resulting in several different parameters such as sound absorption coefficient, sound transmission loss, airflow resistivity, sound power ratio and comparative analysis.

In terms of ASTM and ISO standards, acoustical absorption could be measured by finding and calculating the normal incidence sound absorption coefficient, the random incidence sound absorption coefficient, the sound transmission loss, and the airflow resistivity. All these specific parameters define the acoustical absorption properties of textiles. Table 1 shows the ASTM and ISO standards, which are related to measuring acoustical properties of textiles.

The test standards that are used to measure acoustical properties of nonwoven fabrics directly are shown in Table 1. ASTM 2611-09 standard test method [7] measures the normal incidence sound transmission using a tube, four microphones and a digital frequency analysis system. Normal incidence implies the sound application direction to the specimen as perpendicular direction. This method does not need big reverberation rooms. This method can provide comparison data for small specimens, something that cannot be done in the reverberant room method [7].

ASTM E90-09 test method measures the airborne sound transmission loss of building partitions such as walls of all kinds, operable partitions, floor–ceiling assemblies, doors, windows, roofs, panels and other space-dividing elements [8].

ISO 140 has 14 different parts of acoustical measurement standards for buildings and building elements. The basic standards ISO 140-4 and ISO 140-7 specify the measurement procedure in detail under ideal conditions, but they do not give much information about the specifics and the room properties. The guidelines for the very large reverberation rooms, long and narrow rooms, staircases, coupled rooms, etc., are given in ISO 140-14:2004 standard [9].

ASTM E492 test standard deals with the sound, which comes from the floor. For that reason, specific tapping machine is designed to tap the floor at specific positions. It measures the impact sound transmission of floor–ceiling assemblies using a standardized tapping machine. The assumption is that the test specimen constitutes the primary sound transmission path into a receiving room located directly below so that an approximation to a diffuse sound field exists in this room [10].

Table 1 Standard test methods on acoustical properties of textile materials

Test methods	Title of the method	Measured parameter
ASTM C384-04	Standard test method for impedance and absorption of acoustical materials by impedance tube method	Normal incidence sound absorption coefficient
ISO 10534-1 ISO 10534-2	Determination of sound absorption coefficient and impedance part 1: impedance tube method Determination of sound absorption coefficient and impedance in impedance tubes-part 2: transfer function method	
ASTM E1050-12	Standard test method for impedance and absorption of acoustical materials using a tube, two microphones and a digital frequency analysis system	
ASTM 2611-09	Standard test method for measurement of normal incidence sound transmission of acoustical materials based on the transfer matrix method	
ASTM C423-09	Standard test method for sound absorption and sound absorption coefficients by the reverberation room method	Random incidence sound absorption coefficient
ISO 354	Measurement of sound absorption in a reverberation room	
ASTM E90-09	Standard test method for laboratory measurement of airborne sound transmission loss of building partitions and elements	Sound transmission loss
ISO 140	Measurement of sound insulation in buildings and building elements	
ASTM E492-09	Standard test method for laboratory measurement of impact sound transmission through floor–ceiling assemblies using the tapping machine	Noise reduction and sound transmission loss
ASTM E336-16	Standard test method for measurement of airborne sound attenuation between rooms in buildings	
SAE J1400	Laboratory measurement of the airborne sound barrier performance of flat materials and assemblies	
ASTM C522	Standard test method for airflow resistance of acoustical materials	Airflow resistivity
ASTM E477	Standard test method for laboratory measurements of acoustical and airflow performance of duct liner materials and prefabricated silencers	

ASTM E336-14 standard deals with the sound isolation between two spaces in a building, which is determined by a combination of the direct transmission through the nominally separating building element and any transmission along a number of indirect paths. This standard defines procedures and metrics to assess the sound isolation between two rooms or partitions in a building and it measured the noise reduction (NR) between the rooms [11].

SAE J1400 standard is used for determining the airborne sound barrier performance of materials and composite assemblies commonly installed in surface vehicles and marine products. It provides sound transmission loss measurement of barrier materials. During the test, the transmission loss (TL) is projected from the measured noise reduction of the test sample using a correlation factor (CF). This test method makes the evaluation of automotive materials and assemblies possible under the test conditions of representative size, edge constraint and sound incidence in order to compare the sound properties of specimens [12].

ASTM E477 standard is used for the measurement of acoustical properties of sound attenuating devices including duct liner materials, integral ducts and in-duct absorptive straight and elbow silencers used in the ventilation systems of buildings. In terms of parameters, with this method, acoustical insertion loss, airflow generated noise and pressure drop as a function of airflow are measured [13].

Some other test standard methods, which do not measure the acoustical properties of nonwoven fabrics but measure parameters related to the acoustical properties, are shown in Table 2. ASTM C634 standard is directly related to the terminology of acoustical absorption. ISO 11654, ASTM E1332, ASTM E413 and ISO 717 standards are related on the ratings of the acoustical parameters including sound absorption, sound attenuation and sound insulation. EN TS ISO 140-4 and EN TS 2382 test standards are the Turkish Standard Test Method that is very similar to 1st, 4th and 5th parts of ISO 140.

Table 2 Standard test methods related to acoustical properties of materials

Test methods	Title of the method
ISO 11654:1997	Sound absorbers for use in buildings—rating of sound absorption
EN TS ISO 140-4	Acoustics measurement of sound insulation on buildings and of building elements part 4: field measurements of air borne sound insulation between rooms
EN TS 2382	Field and laboratory measurements of airborne and impact sound transmission
ISO 16283-2:2015	Field measurement of sound insulation in buildings and of building elements—part 2: impact sound insulation
ASTM C634	Standard terminology relating to building and environmental acoustics
ASTM E596	Test method for laboratory measurement of noise reduction of sound-isolating enclosures
ASTM E1332	Classification for rating outdoor–indoor sound attenuation
ASTM E1414	Test method for airborne sound attenuation between rooms sharing a common ceiling plenum
ASTM E413-10	Classification for rating sound insulation
ISO 717	Rating of sound insulation for dwellings
ANSI/ASA S1.11	Octave-band and fractional-octave-band analogue and digital filters

ASTM E1414 standard method uses two horizontally adjacent small rooms for the measurements. The small room partition either extends to the underside of a common plenum space or penetrates through it [14].

ISO 16283-2:2015 standard method deals with the impact sound insulation. The impact source is taken from the floor or the stairs in a building. This method uses the room sizes between 10 and 250 m^3, and the frequency range is used between 50 and 5000 Hz. The test results can be used to measure impact sound insulation in unfurnished or furnished rooms [15].

As seen in Table 1, several different ASTM or ISO standards could be found to measure acoustical properties of textile materials. Normal incidence sound absorption coefficient is measured using small test devices and small sample sizes (as small as 1 cm in diameter) but the test methods used for measuring this parameter cannot measure the acoustical absorption of textile materials at frequencies lower than 100 Hz. On the contrary, random incidence sound absorption coefficient is measured using big reverberation rooms and bigger sample sizes (as big as 12 m^2), but the test methods used for measuring this parameter can measure the acoustical absorption of textile materials at frequencies as low as 20 Hz.

ANSI/ASA S1.11 standard is about octave-band and digital filters. Octave bands are important and they are used for the measurements of acoustical properties of nonwoven fabrics. Octave is a range of frequencies whose upper frequency limit is twice that of its lower frequency limit. For example, the 1000 Hz octave band contains noise energy at all frequencies from 707 to 1414 Hz. 1/1, 1/3, 1/6, 1/12 and 1/24 octaves are all used in acoustics [16, 17].

Sound absorption coefficients are normally reported in a series of number of 1/3-octave-band centre frequencies between 100 and 5000 Hz. In the literature, these frequencies are specifically chosen as the six required standard octave-band centre frequencies of 125, 250, 500, 1000, 2000 and 4000 Hz.

In acoustical measurements, octave bands and the centre frequencies of these bands are defined by ISO—31.5, 63, 125, 250, 500 Hz, 1, 2, 4, 8, 16 kHz to divide the audio spectrum into 10 equal parts are used for sound pressure level and sound absorption coefficients. When additional 1/3-octave-band sound absorption coefficients are needed for the measurements, they can be chosen but they are not considered as "valid" according to ASTM C423 standard.

ASTM E596 standard uses reverberation rooms for the measurement of the noise reduction of sound-isolating enclosures [18]. Most of the acoustical test methods which use reverberation rooms use the Sabin's formula. Wallace Clement Sabine was a pioneer in architectural acoustics. He investigated the impact of absorption on the reverberation time. He then discovered the type of relation between these quantities on October 29, 1898. Sabine derived an expression for the duration, T, of the residual sound to decay below the audible intensity, starting from a 1,000,000 times higher initial intensity as in the following equation:

$$T = 0.161\frac{V}{A}, \tag{2}$$

where V is the room volume in cubic metres (m^3), and A is the total absorption in square metres (m^2).

Sabine's reverberation formula has been applied successfully for many years to determine material absorption coefficients by the means of reverberation rooms. Sabine's formula is still widely accepted as a very useful estimation method for the reverberation time in rooms. Sabin is a unit of Sound Absorption in square metre, referring to the area of open window. This unit stems from the fact that sound energy travelling toward an open window in a room will not be reflected at all, but completely disappears in the open air outside. The effect would be the same if the open window would be replaced with 100 % absorbing material of the same dimensions.

The Noise Reduction Coefficient (NRC) is a single number rating and it is calculated by determining the arithmetic mean of the absorption coefficients in the 250, 500, 1000 and 2000 Hz 1/3-octave frequency bands. This number is rounded to the nearest multiple of 0.05. When the perfect reflection happens, NRC becomes 0 (zero). When the perfect absorption happens, NRC becomes 1 (one). This parameter can also be measured using reverberation room techniques. The noise absorption coefficients of nonwoven fabrics are commonly determined using ASTM C423 standard test method [19]. In ASTM C423-09a, NRC is replaced by Sound Absorption Average (SAA). SAA is very similar to NRC. The only difference is that the nearest SAA number is rounded to 0.01. This is the average of the absorption coefficients for the twelve 1/3-octave bands from 200 to 2500 Hz rather than six 1/3-octave bands from 250 to 2000 Hz used in NRC measurements [20].

Sound classification according to ISO 11654:1997 standard, specification of a method where the frequency-dependent sound absorption can be recalculated into a summary value is described. This summary value can lead to a classification of sound class to compare different types of sound absorbers. Before this can be done, the 1/3 octave-band measurements are converted into octave band according to standard ISO 354:2003 (ISO 11654:1997). To get the value for what sound class the tested object belongs to, it has to be compared to a weighting curve. The weighting curve will from here on be referred to the reference curve. The reference curve starts from 0.8 at 250 Hz and goes up to 1.0 for 500, 1000 and 2000 Hz, and at 4000 Hz it goes down to 0.8 again [21].

Surface impedance is another important parameter for acoustical properties of nonwoven materials. Impedance tube that is used for measurement of sound absorption coefficient works using surface impedance. Surface impedance can be defined as the resistance of the material to the sound wave that interacts with the material. Therefore surface characteristics of the material play an important role. If the effect of surface area is neglected, the surface can be described by specific acoustic impedance as the ratio of pressure to particle velocity at the surface [4].

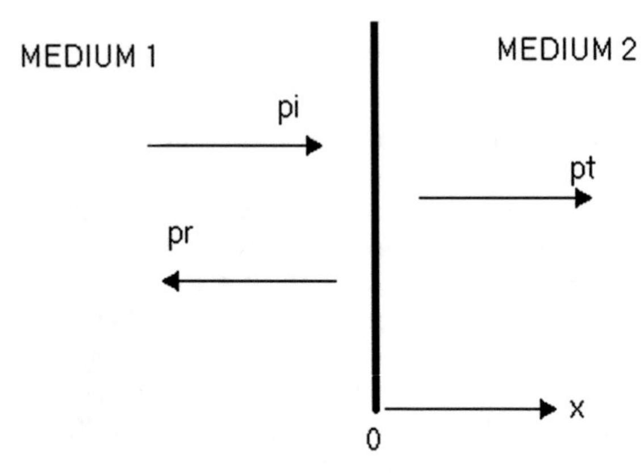

Fig. 2 Sound pressure changes in boundary [22]

The total pressure at medium 1 that is shown in Fig. 2 is calculated from the following equations:

$$p = p_{\rm i} + p_{\rm r} = A e^{j(\omega t - kx)} + B e^{j(\omega t + kx)} \tag{3}$$

$$u = \frac{p_{\rm i} - p_{\rm r}}{z_0}$$
$$z_s = \left(\frac{p}{u}\right)_{x=0} = \frac{A+B}{A-B} z_0 \tag{4}$$

where p is total pressure, $p_{\rm i}$ is incident-wave pressure, $p_{\rm r}$ is reflected wave pressure, A, B are constants, $z_{\rm s}$ is surface impedance, z_0 is characteristic impedance, u is particle velocity, ω is wavelength, k is wave number, x is travelling distance and t is time.

When a new constant is defined as $r = B/A$, Eq. (4) becomes as the following equation:

$$z_{\rm s} = \frac{1+r}{1-r} z_0$$
$$r = \frac{z_{\rm s} - z_0}{z_{\rm s} + z_0} \tag{5}$$

where $z_{\rm s}$ is surface impedance, z_0 is characteristic impedance, r is constant and A, B are the constants from pressure components.

If the incident angle is taken into account, then r will be calculated from the following equation:

$$r = \frac{z_{\rm s} \cos\theta - z_0}{z_{\rm s} \cos\theta + z_0} \tag{6}$$

where $z_{\rm s}$ is surface impedance, z_0 is characteristic impedance, r is constant and θ is the angle of incidence.

From the new constant r, the absorption coefficient can be defined and calculated from the following equation:

$$\alpha_\theta = 1 - |r|^2 \tag{7}$$

where α_θ is absorption coefficient, and r is constant [4, 23].

1.4.1 Acoustical Absorption

One of the most important parameters used for defining acoustical absorption is sound absorption coefficient. Sound absorption coefficient is measured using an impedance tube. The standard method used to measure sound absorption coefficient is the ASTM C384, namely "Standard Test Method for Impedance and Absorption of Acoustical Materials by Impedance Tube Method" and the ISO 10534 standard, namely, "Determination of Sound Absorption Coefficient and Impedance in Impedance Tubes." This method is the most widely used in both industry and the literature for nonwoven fabrics.

The acoustical impedance properties of a sound absorptive material are related to its physical properties, such as airflow resistance, porosity, elasticity and density. The measurements described in ASTM C384 standard are useful for product development of sound absorptive materials [24].

This method covers the use of an impedance tube for the measurement of impedance ratios and the normal incidence sound absorption coefficients. Impedance ratio is the ratio of the specific normal acoustic impedance at the surface to the characteristic impedance of the specimen. The real and imaginary components are called resistance ratio and reactance ratio. Normal incidence sound absorption coefficient of a surface is the fraction of the perpendicularly incident sound power absorption.

A plane wave travels in one direction down a tube. Then it is reflected back by the test specimen in order to produce a standing wave that can be caught with a microphone. The normal incidence sound absorption coefficient is determined from this standing wave ratio at the face of the test specimen. A measurement of a position of the standing wave with reference to the face of the specimen is needed in order to determine the impedance ratio.

Impedance ratio and absorption coefficient are the functions of frequency. These properties are measured with the tones at a number of frequencies chosen. The acoustical impedance properties of sound absorptive material are related to its physical properties, such as airflow resistance, porosity, elasticity and density.

Normal incidence sound absorption coefficients are more useful than random incidence coefficients [24].

The apparatus, which is shown in Fig. 3 is essentially a tube with a test specimen at one end and a loudspeaker at the other. A probe microphone that can be moved along the length of the tube is used to catch the standing wave in the tube. The signal from the microphone is filtered, amplified and recorded.

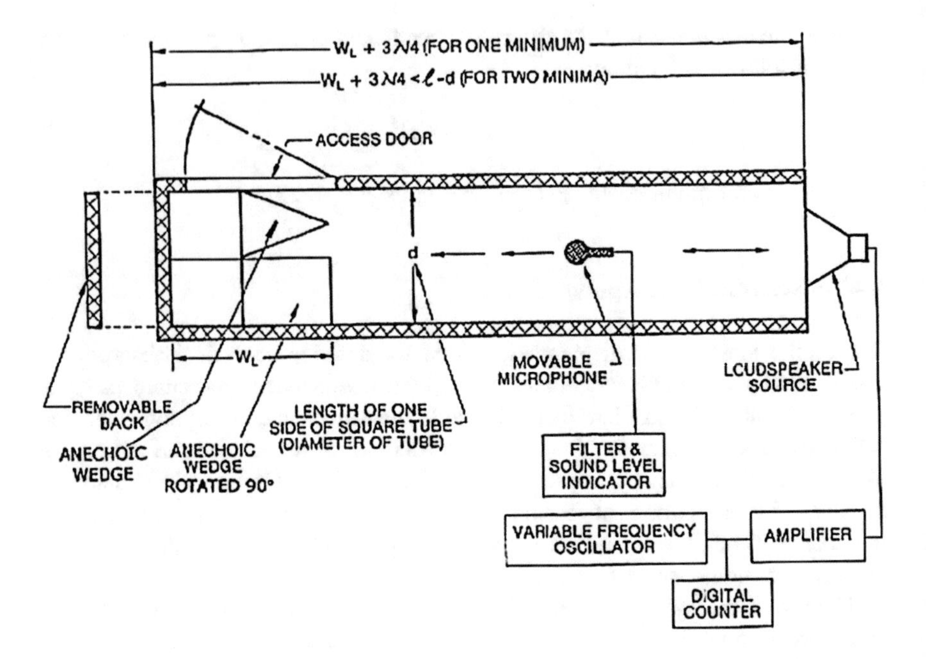

Fig. 3 Typical impedance tube and sound absorption coefficient measurement units [24]

The tube may be made of metal, plastic, Portland cement, or other suitable material that has low sound absorption. Its interior cross section may be circular or rectangular but must be uniform from end to end. The tube must be straight and its inner surface must be smooth, non-porous and free of dust to keep the sound losses low. The interior of the tube may be sealed with paint, epoxy, or other coating material to ensure low sound absorption of the interior surface. The tube walls must be massive and rigid enough so that the propagation of sound energy through them by vibration is negligible [24].

For circular tubes, the upper limit of frequency is

$$f < 0.586 \frac{c_0}{d_t} \tag{8}$$

where f is the frequency of the sound wave, c_0 is the speed of sound in the tube, and d_t is the diameter of the tube.

For rectangular tubes, d_t is used as a symbol for the larger cross-section dimension, and the upper limit is shown at the following equation:

$$f < 0.5 \frac{c_0}{d_t} \tag{9}$$

It is the best to work well below these limits whether the tube is circular or rectangular. Above these frequencies, cross modes may develop, and the incident and reflected waves in the tube are not likely to be plane waves. If sound with a frequency below the limiting value enters the tube as a non-plane wave, it will become a plane wave after travelling a short distance. For this reason, no measurement should be made closer than one tube diameter to the source end of the tube [24].

The length of the tube is also related to the frequency. The tube must be long enough to contain that part of the standing wave pattern needed for measurement. It must be long enough to contain at least one and preferably two sound pressure minima. To ensure that at least two minima can be observed in the tube, its length should be such that

$$f < 0.75 \frac{c_0}{(l - d_t)} \tag{10}$$

where l = Length of tube.

If the tube is 1 m in length and 0.1 m in diameter and the speed of sound is 343 m/s, the frequency should exceed 286 Hz if two sound pressure minima are to be observed.

Generally, the microphone is put on the outside and connected to a hollow probe tube that is inserted through the source end of the apparatus and is aligned with the central axis of the tube. If the microphone is within the tube, then the cross-sectional area of the microphone should be less than 5 % of the cross-sectional area of the tube.

The microphone output should be filtered to remove any harmonics and to reduce the adverse effect of ambient noise. The signal-to-noise ratio of the measuring amplifier must be at least 50 dB [24].

The sound reflection coefficient could also be measured using the test apparatus that has been designed and verified following the ASTM E1050 standard, namely, "Standard Test Method for Impedance and Absorption of Acoustical Materials Using A Tube, Two Microphones and A Digital Frequency Analysis System," to guarantee the reliability of the results in the frequency range of interest as shown in Fig. 4.

A specimen of the porous material of thickness has been placed at the rigid closed end of a straight cylindrical pipe in which a loudspeaker introduces plane sound waves. Defining as $h = p_2/p_1$ the measured frequency response function between the two microphones, it can be proven that the experimental reflection coefficient is given by the following equation:

$$r = \frac{e^{jk2z_1}\left(e^{jks} - h\right)}{h - e^{-jks}} \tag{11}$$

Measurement of sound absorption coefficients of nonwoven fabrics for very low frequencies is possible using big reverberation rooms (as big as 200 m^3). Since low frequency sound waves have very big wavelengths, the absorption of these waves could be measured in a very big area. The known standard for this measurement is

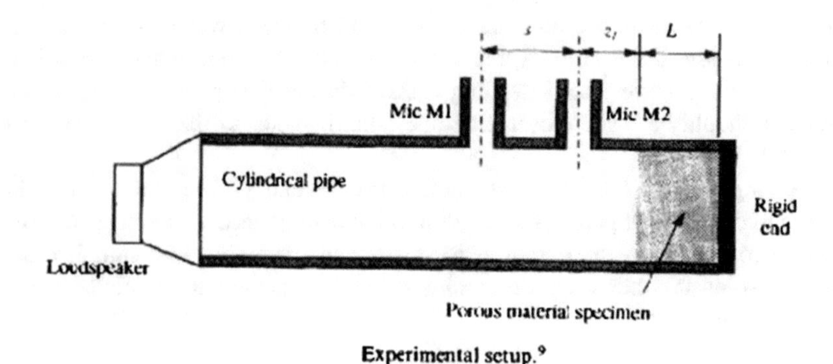

Fig. 4 Testing device that measures sound reflection coefficient based on ASTM 1050 standard [25]

ISO 354:1985, namely, "Measurement of Sound Absorption in a Reverberation Room." ASTM 423-09 standard test method also measures the sound absorption using reverberation rooms. This test method is titled as "Standard Test Method for Sound Absorption and Sound Absorption Coefficients by the Reverberation Room Method."

In ASTM C423 test method [19], the absorption of sound in a room is operationally defined by the Sabine equation

$$A = 0.921 \frac{V \times d}{c} \tag{12}$$

where A is the room absorption in metric Sabin, V is the volume of the room in m^3, d is the decay of sound pressure level in the room in dB/s, and c is the speed of sound in m/s.

The speed of sound is calculated using the following equation:

$$c = 20.047\sqrt{(273 + T)} \tag{13}$$

where c is the speed of sound in m/s, and T is the temperature of the environment in °C.

In this measurement technique, first the sound absorption is measured when the room is empty and then sound absorption is measured again when the nonwoven fabric is in the room. Then the difference is divided by the area of the specimen (nonwoven fabric), sound absorption coefficient is calculated using the following equation:

$$\alpha = \frac{A_2 - A_1}{S} \tag{14}$$

where α is the absorption coefficient of specimen (dimensionless) absorption of the room after the specimen has been brought in, S is the area of the specimen, A_1 is the absorption of the reverberation room without specimen in it, and A_2 is the absorption of the reverberation room with a specimen in it.

The speed of the sound wave could be measured using Kundt's tube. This technique is very old technique discovered by German physicist August Kundt [26]. Kundt's tube is a horizontal transparent tube with powder such as talc inside. When the loudspeaker on one side generates the single wave (pure tone) sound into the tube, the powder inside shapes as the line on every wavelength of the sound wave. The distance between the lines or the piles of the talc powder shows the wavelength of the sound [27, 28].

Sound absorption coefficients normally range from a numerical value of zero to one and are defined, ideally, as the fraction of random-incident sound power absorbed by the specimen. Sometimes values are reported outside this coefficient range. Coefficients just below a value of zero may occur because of the uncertainty of the measurement. Coefficients exceeding a value of 1 may occur for the same reason or due to the diffraction effects.

1.4.2 Sound Transmission Loss (STL)

Sound transmission loss is an important parameter that shows how much nonwoven fabric transmits sound waves and how much sound energy is lost in the fabric by energy change. The sound transmission loss could be measured using both impedance tubes and reverberation room techniques. In the procedure, the sound wave is sent to the nonwoven fabric, and the microphone on the other side measures the amount of sound energy. Since the sound energy that was sent at the first place is known, the difference between this value and the one measured by the microphone calculates the sound transmission loss.

The sound transmission loss values are expressed to statistical precision according to the equation

$$\mathrm{STL} = L_1 - L_2 + 10\log_{10}(A_s/A_e) \tag{15}$$

where STL is sound transmission loss, L_1 and L_2 are the space averaged sound pressure levels, A_s is the area of the specimen, and A_e is the equivalent absorption area [4].

From the measured STL values, the sound transmission procedure is specified in the ASTM Standard E413-87.

Sound transmission loss can also be calculated using the following equation:

$$\mathrm{STL} = 10\log\left(\frac{1}{\tau}\right) \tag{16}$$

where τ is the transmission coefficient and it is defined as the ratio of intensity of transmitted wave to intensity of the incident wave.

1.4.3 Airflow Resistivity (AFR)

In order for a material to absorb sound, it must be porous, so that the sound waves can move into the material. Sound propagates through narrow channels of the material and dissipates the energy either by loss due to friction between fibres, viscous dissipation, or a combination of both. Thus, the porosity of the material has a significant effect on sound absorption performance. A very porous material cannot generate enough energy loss to provide high absorption because the sound wave goes through the material without touching the fibres. A material with low porosity may not allow sound to propagate into the material to get absorbed.

Flow resistance is another important term for understanding the sound absorption properties of textiles. A highly porous material has a very low flow resistance, and vice versa. Therefore, for optimum sound absorption performance, an optimum value of the flow resistance needs to be attained. The flow resistance characteristics of a textile material depend on

1. the bulk volume density of the fibre;
2. the volume density of the fibre;
3. the diameter of the fibre;
4. the elasticity modulus of the fibre, and the internal damping of the fibre.

Fibres and their interlocked airspaces are the frictional elements that provide resistance to the sound wave motion. As the sound wave goes through the material, its amplitude is decreased by friction as the wave attempts to move through the passages. Therefore the energy of the sound wave decreases [29].

This test method (ASTM C522 standard) covers the measurement of airflow resistance, specific airflow resistance, and airflow resistivity of porous materials that can be used for the absorption of sound. Materials cover a range from thick boards or blankets to thin mats, fabrics, papers and screens. When the material is anisotropic, provision is made for measurements along different axes of the specimen.

This test method is designed for the measurement of values of specific airflow resistance ranging from 100 to 10,000 rayls (Pa s/m) with linear airflow velocities ranging from 0.5 to 50 mm/s and pressure differences across the specimen ranging from 0.1 to 250 Pa.

This test method describes the measurement of a steady flow of air through a specimen, the air pressure difference across the specimen, and the volume velocity of airflow through the specimen.

Valid measurements are made only in the region of laminar airflow where, aside from random measurement errors, the airflow resistance is constant. When the airflow is turbulent, the apparent airflow resistance increases with an increase of volume velocity and the term, "airflow resistance," does not apply [29].

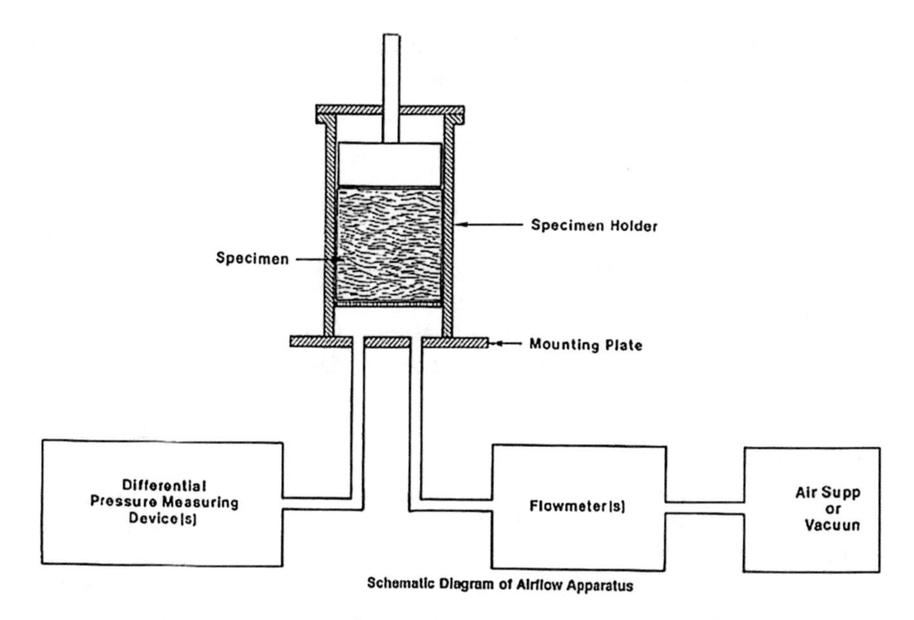

Fig. 5 An apparatus for the airflow measurement [30]

The apparatus for the airflow measurement is shown in Fig. 5. Air supply is a suction generator or positive air supply arranged to draw or force air at a uniform rate through the test specimen. A flowmeter is used to measure the volume velocity of airflow through the specimen. It is better to have two or more flow meters with overlapping ranges to enable different airflow velocities to be measured to the same precision.

A differential pressure measuring device measures the static pressure difference between the faces of the specimen with respect to atmosphere.

The test specimen is mounted according to the type of test to be done. Then the specimen holder is sealed to the mounting plate and the airflow is adjusted to give readable settings on the flowmeter and pressure measuring device. An airflow velocity is started well below 50 mm/s. After that, the differential pressure is recorded as the flow rate. The measurements are repeated several times, using a larger airflow rate each time. Finally, the parameter that is required is calculated.

$$R = \frac{P}{U_{\mathrm{f}}} \tag{17}$$

$$R_{\mathrm{s}} = \frac{A_{\mathrm{s}}P}{U_{\mathrm{f}}} \tag{18}$$

$$r_0 = \frac{A_{\mathrm{s}}P}{LU_{\mathrm{f}}} \tag{19}$$

where R is airflow resistance, R_s is specific airflow resistance, P is differential pressure, U_f is flow rate, A_s is the area of specimen and L is the thickness of specimen [30].

1.4.4 Comparative Sound Absorption Measurement

Sound absorption of nonwoven fabric could be compared without using very complicated testing devices. One of the examples for comparative measurement of sound absorption is called the Clemson–Boston Differential Sound Insulation Tester [7, 23–26]. As shown in Fig. 6, this testing device consists of six components: (1) the sound signal processing computer; (2) the sound signal amplifier; (3) the sound source; (4) the sound chamber; (5) the sound detector; and (6) the material sample holder [22].

The signal processing computer (1) generates the sound signals. The signals are amplified by the signal amplifier (2). The amplified signals are sent to the sound source (3) to be converted to sound waves. The material to be tested is mounted in a sample holder (6) and placed in the sound path. The sound wave is transmitted inside the sound chamber (4) passing through the sample material to the sound detector (5). The sound detector converts the received sound signals to electrical signals and they are sent back to the signal processing computer to be analyzed. If no sample is in the path of the sound wave, a background reference is obtained and sample data can be compared to this air reference. If two samples are compared, data from one sample is compared to the other with no changes to the settings of the instrument [31–34].

This sound spectrometer is not the perfect sound chamber with perfect acoustical properties, but it gives a comparison of the samples under the same conditions and therefore yields a direct acoustical comparison.

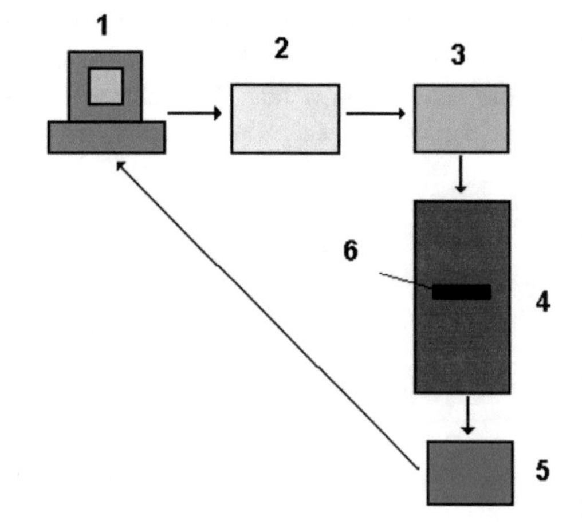

Fig. 6 Components of the Clemson–Boston Differential Sound Insulation Tester [22]. (*1*) Computer, (*2*) amplifier, (*3*) sound source, (*4*) sound chamber, (*5*) sound detector, (*6*) sample holder

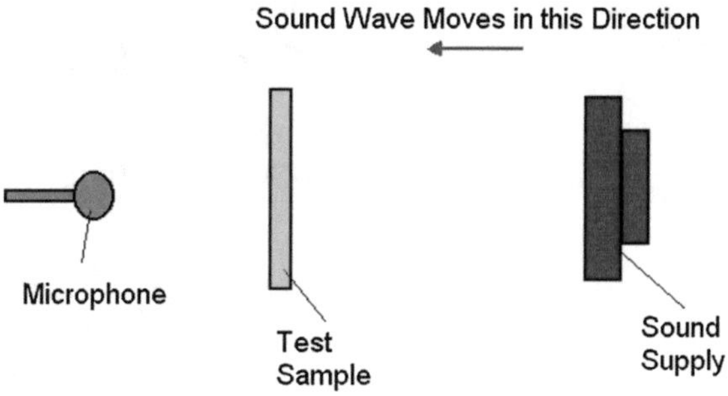

Fig. 7 The working principle of the Clemson–Boston Differential Sound Insulation Tester [22]

The design of the sound spectrometer takes in several other important considerations. The distance between the sound detector and the sound source is 1 m. This distance was used because some audio standards use the one-metre distance to measure the sound pressure level of speakers. The holder is mounted on a track allowing the sample to move for variable distances between the source and detector. This was done because the sound pressure varies with the distance from the source and will give more detailed and varied analysis. The sample size is 12 in. by 12 in. This makes it more convenient to compute and compare samples by the square foot. This also gives a comparison of the fibre arrangement, thickness, surface and the other finished details. Two sound detectors are used: one in front of the sample and one behind the sample. Therefore, transmitted and reflected sound can be selected and measured as a comparison. The sound spectrometer can be used to measure the sound insulation difference between two materials. Although there is a special microphone for reflected sound, this sound chamber is not capable of measuring reflected sound. This device tests only transmitted sound.

The working principle of the Clemson–Boston Differential Sound Insulation Tester is shown in Fig. 7. A sound supply program generates a sound signal. The sound waves, which are in different frequencies, travel to the test sample. When the sound waves hit the test sample, some are transmitted, some are reflected, and the rest are absorbed by the test sample. The microphone senses the transmitted waves, and these sound waves are amplified and sent to a computer [22, 31–34].

1.4.5 Calculation of Sound Power Ration Between Two Samples

Although decibel values are measured for the samples, the calculation of the sound power ratio between two samples is possible. First, a test without the sample is done and then the result is taken as a reference value. Based on this reference value, the sample is put to the sample holder and the test is done again. This test is taken as the resulting value for Sample 1. Then the same procedure is followed for Sample 2. Finally, the sound transmission difference between two samples is calculated from

$$D_1 - D_s = 10\log\frac{P_{s1}}{P_a} - 10\log\frac{P_{st}}{P_a} = 10\log\frac{P_{s1}}{P_{st}} \quad (20)$$

$$D_2 - D_s = 10\log\frac{P_{s2}}{P_a} - 10\log\frac{P_{st}}{P_a} = 10\log\frac{P_{s2}}{P_{st}} \quad (21)$$

$$(D_1 - D_s) - (D_2 - D_s) = 10\log\frac{P_{s1}}{P_{st}} - 10\log\frac{P_{s2}}{P_{st}} = 10\log\frac{P_{s1}}{P_{s2}} \quad (22)$$

$$\frac{P_{s1}}{P_{s2}} = 10^{\frac{(D_1 - D_2)}{10}} \quad (23)$$

where D_1, D_2 and D_s are the decibel measurements of Sample 1, Sample 2 and without a sample, respectively. P_{s1}, P_{s2} and P_{st} are the sound power values measured by the microphone for Sample 1, Sample 2 and without a sample, respectively. P_a is the sound power generated from the microphone.

Equations 20–23 prove that the sound power ratio of two samples can be calculated directly from measured decibel values. This ratio may be related to the number of fibres in nonwoven fabric and total maximum surface area in nonwoven fabric.

Other than the test methods directly used for measuring acoustical absorption of textile materials, some other ASTM, ANSI and ISO test standards could help adjust the direct measurement techniques. Some of these methods could be listed as the following: ASTM C413 (Classification for Rating Sound Insulation); ANSI S12.31 (American National Standard Precision Methods for the Determination of Sound Power Levels of Broad-Band Noise Sources in Reverberation Rooms); ANSI S12.32 (American National Standard Precision Methods for the Determination of Sound Power Levels of Broad-Band Noise Sources in Reverberation Rooms); and ISO 3741 (Determination of sound power levels and sound energy levels of noise sources using sound pressure-Precision methods for reverberation test rooms).

1.5 Conclusion

Evaluation methods for acoustical nonwoven fabrics are discussed in this chapter. The acoustical performance of the fabrics could only be measured and evaluated using these methods and procedures given in this chapter. This chapter dealt with only the most important physical and mechanical properties of nonwoven fabrics that are directly related to the acoustical properties. The measurements techniques, test methods and standards are also explained more in detail.

Mechanical properties of nonwoven fabrics that are related to acoustical properties are tensile properties and Young's modulus. Tensile strength is defined as the maximum tensile stress a material can withstand before it fails. The measurement of tensile strength is done using UTM. The grab tensile tests are done according to the

ASTM D5034 (grab method). Young's modulus is defined as the ratio of linear stress to linear strain. It is also measured using a UTM, and Young's modulus is calculated by dividing stress to strain in the elastic region.

Acoustical absorption properties of nonwoven fabrics are determined using sound absorption coefficient, sound transmission loss, noise reduction coefficient, sound impedance and airflow resistivity parameters. These parameters could be measured using many different test methods. These methods are explained and related ASTM, ISO, ANSI and SAE test standards are discussed in this chapter. Especially impedance tubes and reverberation rooms are used for most of the acoustical test methods discussed in this chapter.

Several different ASTM or ISO standards could be found to measure acoustical properties of textile materials. Normal incidence sound absorption coefficient is measured using small test devices and small sample sizes (as small as 1 cm in diameter) but the test methods used for measuring this parameter cannot measure the acoustical absorption of textile materials at frequencies lower than 100 Hz. On the contrary, random incidence sound absorption coefficient is measured using big reverberation rooms.

The Noise Reduction Coefficient (NRC) is a single number rating and it is calculated by determining the arithmetic mean of the absorption coefficients in the 250, 500, 1000 and 2000 Hz 1/3-octave frequency bands.

Surface impedance is another important parameter for acoustical properties of nonwoven materials. Impedance tube that is used for the measurement of sound absorption coefficient works using surface impedance.

Sound absorption coefficient is measured using an impedance tube. The standard method used to measure sound absorption coefficient is the ASTM C384. This method covers the use of an impedance tube for the measurement of impedance ratios and the normal incidence sound absorption coefficients. Impedance ratio is the ratio of the specific normal acoustic impedance at the surface to the characteristic impedance of the specimen.

Sound transmission loss is an important parameter that shows how much nonwoven fabric transmits sound waves and how much sound energy is lost in the fabric by energy change. The sound transmission loss could be measured using both impedance tubes and reverberation room techniques. In the procedure, the sound wave is sent to the nonwoven fabric, and the microphone on the other side measures the amount of sound energy. Since the sound energy that was sent at the first place is known, the difference between this value and the one measured by the microphone calculates the sound transmission loss.

Flow resistance is another important term for understanding the sound absorption properties of textiles. A highly porous material has a very low flow resistance, and vice versa. Therefore, for optimum sound absorption performance, an optimum value of the flow resistance needs to be attained.

References

1. Glé et al (2011) Acoustical properties of materials made of vegetable particles with several scales of porosity. Appl Acoust 249–259
2. Nayak R, Punj S, Chatterjee K, Behera B (2009) Comfort properties of suiting fabrics. Indian J Fibre Text Res 34:122–128
3. Gerges SNY, Balvedi AM (1999) Numerical simulation and experimental tests of multilayer systems with porous materials. Appl Acoust 58:403–418
4. Nayak R, Padhye R (2015) Garment manufacturing technology. Elsevier, Amsterdam, pp 59–80
5. Ford RD (1970) Introduction to acoustics. Elseiver Publishing, New York
6. Conrad J, Jr Hemond (1983) Engineering acoustics and control. Prentice Hall Inc., New Jersey
7. Reynolds DD (1981) 'Engineering principles of acoustics', noise and vibration control. Allyn and Bacon Inc., Boston
8. Tascan M, Vaughn EA (2008) Effects of fiber denier, fiber cross-sectional shape and fabric density on acoustical behavior of vertically lapped nonwoven fabrics. J Eng Fibers Fabr 3(2):32–38
9. Standard Test Method for Measurement of Normal Incidence Sound Transmission of Acoustical Materials Based on the Transfer Matrix Method, ASTM 2611-09, American Society of Testing and Materials. http://www.astm.org/Standards/E2611.htm, 12 Jan 2016
10. Acoustics—measurement of sound insulation in buildings and of building elements—Part 14: guidelines for special situations in the field, ISO 140-14:2004, International Organization for Standardization. http://www.iso.org/iso/catalogue_detail.htm?csnumber=31756, 15 Jan 2016
11. Standard Test Method for Laboratory Measurement of Impact Sound Transmission Through Floor-Ceiling Assemblies Using the Tapping Machine. ASTM E492-09, American Society of Testing and Materials. http://www.astm.org/Standards/E492.htm, 12 Jan 2016
12. Standard Test Method for Measurement of Airborne Sound Attenuation between Rooms in Buildings, ASTM E336-16, American Society of Testing and Materials. http://www.astm.org/Standards/E336.htm, 16 Jan 2016
13. Laboratory Measurement of the Airborne Sound Barrier Performance of Flat Materials and Assemblies. SAE J1400, Society of Automotive Engineers International. http://standards.sae.org/j1400_201008/, 15 Jan 2016
14. Standard Test Method for Laboratory Measurements of Acoustical and Airflow Performance of Duct Liner Materials and Prefabricated Silencers, ASTM E477-13e1, American Society of Testing and Materials. http://www.astm.org/Standards/E477.htm, 17 Jan 2016
15. Standard Test Method for Airborne Sound Attenuation Between Rooms Sharing a Common Ceiling Plenum, ASTM E1414, American Society of Testing and Materials. http://www.astm.org/Standards/E1414.htm, 17 Jan 2016
16. Acoustics—field measurement of sound insulation in buildings and of building elements—Part 2: impact sound insulation, ISO 16283-2:2015, International Organization for Standardization. http://www.iso.org/iso/iso_catalogue/catalogue_tc/catalogue_detail.htm?csnumber=59747, 17 Jan 2016
17. Tannenbaum B, Stillman M (1973) Understanding sound. McGraw-Hill Book, Company, New York
18. Octave-band and Fractional-octave-band Filters—Part 1: specifications (a nationally adopted international standard), ANSI/ASA S1.11 PART 1, American National Standard Electroacoustics. https://global.ihs.com/doc_detail.cfm?&item_s_key=00009495, 17 Jan 2016
19. Standard Test Method for Laboratory Measurement of Noise Reduction of Sound-Isolating Enclosures, ASTM E596-96(2009), American Society of Testing and Materials. http://www.astm.org/Standards/E596.htm, 18 Jan 2016
20. Standard Test Method for Sound Absorption and Sound Absorption Coefficients by the Reverberation Room Method, ASTM C423-09a. Reapproved in 2009, American Society of Testing and Materials. http://www.astm.org/Standards/C423.htm, 10 Jan 2016

21. 'What is NRC, STC and SAA?' http://acoustical.com/what-is-nrc-stc-saa/, 02 Jan 2016
22. Acoustics—sound absorbers for use in buildings—rating of sound absorption, ISO 11654:1997, International Organization for Standardization. http://www.iso.org/iso/catalogue_detail.htm?csnumber=19583, 13 Jan 2016
23. Tascan M (2015) Acoustical properties of fiber network structures. Doctoral thesis, Clemson University (May)
24. Kinsler L, Frey A, Coppens A, Sanders J (2000) Fundamentals of acoustics. Wiley, New York. ISBN 0-471-84789-5
25. Standard Test Method for Impedance and Absorption of Acoustical Materials by the Impedance Tube Method, ASTM C 384-98, 1998, American Society of Testing and Materials. http://www.astm.org/Standards/C384.htm, 08 Jan 2016
26. Braccesi C, Bracciali A (1998) Least squares estimation of main properties of sound absorbing materials through acoustical measurements. Appl Acoust 54(1):59–70
27. Kundt A (January–June 1868). Acoustic experiments. The London, Edinburgh and Dublin Philosophical Magazine and Journal of Science (vol 35, no.4). Taylor & Francis, UK, pp 41–48. Retrieved 2009-06-25
28. Poynting JH, Thomson JJ (1903) A textbook of physics: sound, 3rd edn. Charles Griffin & Co., London, pp 115–117
29. Faber TE (1995) Fluid dynamics for physicists. Cambridge University Press, UK, p 287. ISBN 0-521-42969-2
30. Conrad J, Hemond Jr (1983) Engineering acoustics and control. Prentice Hall Inc., New Jersey
31. Standard Test Method for Airflow Resistance of Acoustical Materials, ASTM C 522-87, American Society of Testing and Materials. Reapproved 1997. http://www.astm.org/Standards/C522.htm, 10 Jan 2016
32. Tascan M, Vaughn EA, Stevens KA, Brown PJB (2011) Effects of total surface area and fabric density on the acoustical behavior of traditional thermal-bonded highloft nonwoven fabrics. J Text Inst 102(9):746–751
33. Tascan M, Vaughn EA (2008) Effects of fiber specific surface and fabric density on the acoustical behavior of needle punched nonwoven fabrics. Text Res J 78:289–296
34. Tascan M, Gaffney KL (2012) Effect of glass beads on sound insulation properties of nonwoven fabrics. J Eng Fibers Fabr 7(1):101–105
35. Standard Test Method for Laboratory Measurement of Airborne Sound Transmission Loss of Building Partitions and Elements, ASTM E90-09, American Society of Testing and Materials. http://www.astm.org/Standards/E90.htm, 12 Jan 2016

Applications of Acoustic Textiles in Automotive/Transportation

Jorge P. Arenas

Abstract This chapter discusses the use of acoustic textiles in the transportation industry; the sector currently represents the most important application of textiles in the world. In general, acoustic textiles are used in the transportation industry to reduce interior noise and vibration and improve the sensation of ride comfort for the passengers. Interior noise is currently a competitive quality characteristic of every mode of transport facility in particular automobiles. Although interior noise lowers the comfort feeling inside a vehicle, it also induces fatigue and may reduce driving safety. The number of cars exceeds by many times the total number of other means of transport produced every year in the world. Therefore, the use and development of acoustic textiles have been more important in the automobile industry mainly because of economic reasons. A variety of sources contribute to the interior noise of a vehicle which can be structure-borne or airborne sound. Acoustic textiles used to control noise in vehicles must provide airborne transmission reduction, damping and sound absorption. However, the use of acoustic textiles in vehicles is not only dependent on their acoustic properties but also on additional characteristics. Selection of a particular material is also determined by its ratio between performance and cost. Acoustic textiles employed to reduce noise and vibrations are used either individually or as components of complex composite materials which are an interesting area of research. Although there are several books that have covered the subject of technical textiles in the transportation industry, this chapter aims to discuss those textile developments that are used mainly to provide noise isolation and sound absorption in different means of transport. This chapter is preceded by an introductory section. Then the use of acoustic textiles for noise control of main sources in cars is discussed. The final section is devoted to further applications of acoustic textiles in aircraft, trains, ships and spacecraft.

Keywords Acoustic textiles · Noise control · Vibration control · Interior noise · Acoustic materials · Sound absorption

---

J.P. Arenas (✉)
Universidad Austral de Chile, Valdivia, Chile
e-mail: jparenas@uach.cl

R. Padhye and R. Nayak (eds.), *Acoustic Textiles*, Textile Science and Clothing Technology, DOI 10.1007/978-981-10-1476-5_7

1 Introduction

Most people relate textiles to the manufacturing of fabrics for clothing and furnishings. However, the use of textiles in the transportation industry currently represents the most important application of textiles in the world. Every year, millions of square metres of textiles are manufactured specifically to satisfy the demands of the transportation sector. Among these, technical textiles employed to reduce noise and vibrations are becoming an interesting area of acoustical research.

A large number of technical textile components are used in modern vehicles. Some of the non-acoustical main applications of technical textiles are seat belts, air bags, filters, padding for sun visors, tyre reinforcement, battery separators, hoses and belts [1, 2]. The term *acoustic textile* has been primarily applied to those materials that have been produced for the specific purpose of providing high values of sound absorption and isolation. The main component of an acoustic textile is fibres that could be either natural or synthetic [3]. Common acoustic textiles used in transportation are woven, knitted and nonwoven [4, 5].

Several studies on the acoustic properties of acoustic textiles have been reported in literature [6–15]. Many studies have also developed theoretical approaches to predict the acoustic performance of sound-absorbing materials from their material properties [16–22].

Active and passive noise control approaches can be adopted for noise control in different means of transport. Active noise control is the process of reducing the total noise using a secondary sound source, whereas passive noise control is provided by sound-absorbing and sound-isolating materials, for example acoustic textiles. Active noise control is achieved by the use of complex digital signal processing techniques to implement a destructive interference of sound waves. The process of active noise control is beyond the scope of this book. Hence, only passive noise control techniques used in cars, aircrafts, ships, trains and other mode of transport facilities will be discussed in the following sections.

2 Automotive Textiles and Vehicle Noise

Although this section is mainly focused on cars, the discussion is also applicable to trucks, buses, motor homes and caravans. The automotive industry is most likely the largest consumer of acoustic textiles. The amount of textiles used for a standard passenger car can be as much as 42 m^2. Acoustic textiles made of polyethylene terephthalate (PET) and nylon fibres are the most used in vehicles to provide sound absorption, although glass fibre is also used for reinforcing hard composite panels to provide sound isolation.

In general, acoustic textiles are used in vehicles to reduce interior noise and vibration and improve the sensation of ride comfort for the vehicle's occupants [4, 23]. Interior noise is currently a competitive quality characteristic of automobiles.

Speech intelligibility measurements are sometimes used for assessing the ability of the customers inside the vehicle to hold a conversation or listen to music during vehicle operation.

The total sound and vibration that a customer feels when he/she operates a vehicle is called harshness. Thus, engineers are often challenged to predict and test the noise, vibration and harshness (NVH) characteristics of a vehicle. In the past, improving the sound quality of a vehicle simply consisted of reducing the overall sound pressure level. Now many vehicle manufacturers are increasingly using subjective studies to better correlate their vehicle quality to customer perceptions through what is called *product sound quality* [24]. This is because the sound quality of a manufactured product involves a psychological judgment. Design engineers are focused on the additional metrics of the vehicle sound such as loudness, frequency content, tonal components, impulsiveness and level fluctuations. It has been observed that the dominant factor in the judgment of the sound quality of a vehicle is its loudness [25]. Sound quality metrics are increasingly used, especially for luxury cars.

On the other hand, most new cars may incorporate high-quality audio systems that are also a valued characteristic of a vehicle product. As in any room, an audio system requires special acoustic conditions inside the vehicle to properly operate. Basically, these acoustical conditions that must be accounted for are (1) low interior noise so the sound reproduced by the car's loudspeakers has the highest possible signal-to-noise ratio at the driver's ear location, (2) a proper reverberation time in the passengers' cabin which depends on the interior total sound absorption, and (3) a relatively uniform interior spatial distribution of the sound so that large undesirable spatial variations are avoided.

Control of the interior frequency response may be achieved by modifying the cabin geometry so that high acoustic resonance peaks are reduced and the interior frequency response tends to be linear. However, the new design of automobile cabin shapes is usually limited, and the acoustic resonances are often damped through the use of sound absorption [26]. Addition of acoustic damping in the form of a sound-absorbing material on the surfaces greatly affects the acoustic character of the passenger space. Figure 1 shows the frequency response of a cabin enclosure before and after adding a sound absorption material. In this example, the addition of a thick nonwoven textile lining significantly reduces the resonant response peaks of the acoustic modes at mid and high frequencies.

2.1 Sources of Noise Within Vehicles

A variety of sources contribute to the interior noise of a vehicle which can be structure-borne or airborne sound. The main sources of noise and vibration in vehicles include those related to the power system (engine, cooling fans, gearboxes and transmissions, brakes and inlet and exhaust systems) and those non-power system sources generated by the vehicle motion (tyre/road interaction noise and

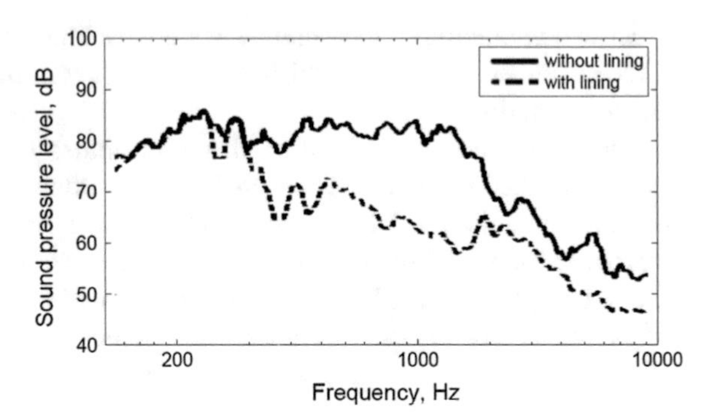

Fig. 1 Frequency response of a cabin enclosure with and without an absorbing lining

aerodynamic noise caused by flow over the vehicles) [27]. In general, structure-borne sound is the dominant source of interior total vehicle noise below 400 Hz, and airborne noise is dominant above 400 Hz. Acoustical coupling between the power train component excitation and the vehicle cabin cavity is also a dominant source of up to 50 Hz of vehicle interior low frequency boom noise [28].

The noise generation mechanisms of the tyre/road noise of a vehicle are vibration related (tread impact and adhesion) and aerodynamic (air displacement as a result of the interaction between the tyre and the road surface) [29]. The level of this noise is also dependent upon vehicle operating conditions and road characteristics that could vary from concrete pavement to different porous asphalt mixes. Power train noise is the dominant noise source during the acceleration of most vehicles. Although power system noise in internal combustion engine cars has been reduced over the years, interior noise of most new cars is dominated by the sources generated by the vehicle motion. This fact is particularly important at speeds over 50 km/h, where tyre/road interaction noise and aerodynamic noise are the dominating sources [29]. Studies have shown that tyre/road noise tends to dominate during cruise conditions on a smooth road at about 80 km/h. This noise increases about 10 dB for each doubling of speed, which represents an approximate doubling of subjective loudness [27]. For hybrid and electric vehicles that use quiet electric motors, the power system noise can almost be neglected in comparison to rolling noise. Therefore, the non-power system noise has received much attention from manufacturers and researchers in recent years.

At higher speeds above about 120 km/h, vehicle noise starts to become dominated by aerodynamic flow noise from flow separation regions at various locations on the vehicle body. The noise generation mechanism is generally due to vortex shedding that creates a complex, unsteady flow impinging on the vehicle body. Major aerodynamic sound sources on the vehicle body include A-pillar flow, flow around exterior rear-view mirrors, windshield wipers, roof racks, radio antenna, underbody and rear-end flow [30].

Since sound power is proportional to the area of a vibrating structure, the car roof also represents an important source of noise to the vehicle interior. Both structure-borne and aerodynamic noises induce vibration of the metal roof resulting in a significant radiating panel-like sound source. This source is important at

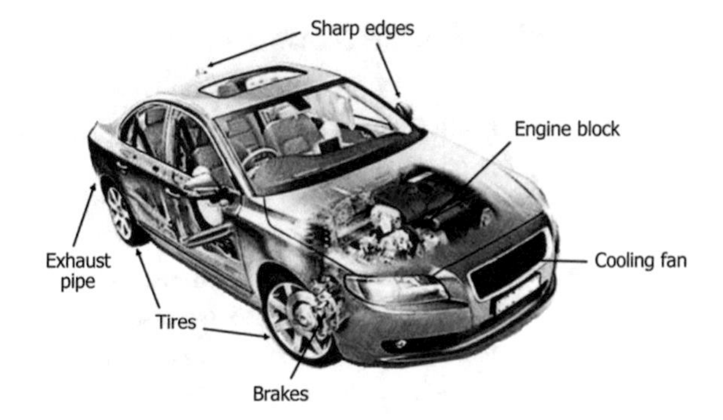

Fig. 2 Location of some of the sources of power system, tyre, and aerodynamic noise on an automobile

frequencies that are close to the frequencies of the first structural modes of the roof [28].

All the sources described above produce low frequency noise, whistling tones that can be transmitted structure-borne into the vehicle interior, and irritating sound modulation that lowers the comfort feeling inside the vehicle, induces fatigue, and may reduce driving safety. Figure 2 shows the location of major sources of noise on an automobile.

Other sources that contribute to the interior noise in vehicles are the components of the heating, ventilating and air conditioning (HVAC) equipment such as the air ducts and vents, the blower, and the manifold. These components need to be examined in detail to reduce their sound contributions. Automotive HVAC noise can be very annoying to the passengers, particularly when it is operated under maximum airflow conditions. Figure 3 shows the in-cabin noise levels of two different cars under idle conditions when the HVAC system is off and on.

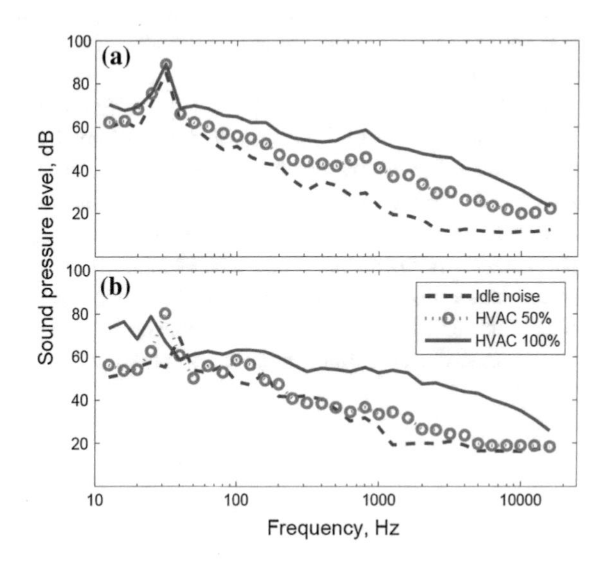

Fig. 3 Sound pressure level inside an automobile produced by the HVAC system at driver's ear location: **a** American sport utility vehicle (SUV) and **b** Japanese compact car

2.2 *Interior Vehicle Noise Control*

Vehicle interior sound pressure levels can be controlled by reducing the noise generated by the sources, by reducing the noise during transmission through airborne, and structure-borne paths, and by reducing the noise transmitted within the vehicle.

Materials used to enclose noise sources are termed *barrier materials* in the automotive industry [28]. The noise isolation performance of these materials is mostly dominated by their mass/unit area. This is a design challenge since car weight reduction is also a requirement of the transportation industry for fuel reduction. In general, barrier materials are characterized by transmission loss (TL), that is, 10× log of the fraction of incident energy that is transmitted. For single layers, TL increases theoretically by 6 dB for each doubling of frequency or by 6 dB at a given frequency if their mass/unit area is doubled. Much better performance can be achieved using multilayer panels, and the TL for such panels can be more like 12 dB/octave rather than 6 dB/octave of a single layer [28].

Noise isolation of the vehicle also depends upon the TL of the vehicle windows. Thus, the use of laminated glass is usually employed to provide impedance mismatch and vibration damping.

Noise reduction is also achieved by providing mechanical damping to the structural vibrating panels of the car body, particularly at resonance frequencies. Constrained and unconstrained viscoelastic layers are typically used for this purpose. Damping layer materials add mass, which can also reduce airborne sound transmission through areas such as floor panels [28].

Sound absorption can also help in reducing interior noise once airborne and structure-borne sound has penetrated into the passenger cabin. TL is combined with the vehicle interior average sound absorption to obtain the total noise reduction. The increase of noise isolation is mathematically estimated by 10× log of the sound absorption, that is, noise reduction increases by 3 dB for each doubling of total sound absorption.

Sound absorption can be provided on the interior surfaces of the vehicle (sidewalls, rooftop and floor) or within the volume by the seats [4, 23, 31]. In buses, the surface below overhead compartments can be used to add extra sound absorption. Although the materials are usually selected for factors other than sound absorption, such as resistance to mechanical damage, ease of cleaning, appearance and acoustic performance, there are several surface areas that can be designed with sound absorption in mind. These include headliners, door casings, carpets, and other interior trims. Figure 4 shows locations where barrier and sound-absorbing materials are often applied in an automobile.

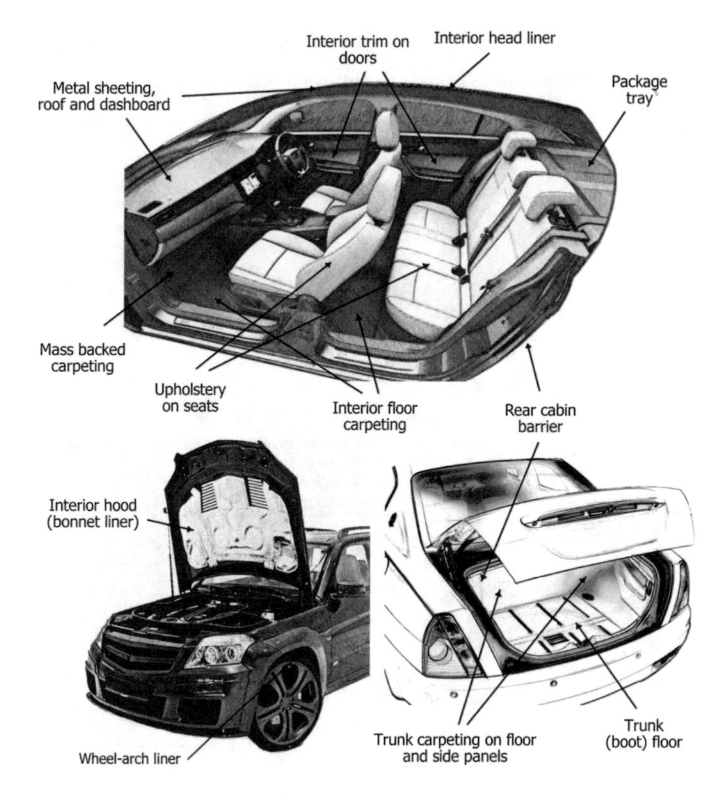

Fig. 4 Typical locations in an automobile where barrier and sound-absorbing materials are utilized

3 Use of Acoustic Textiles in Vehicles

As discussed above, acoustic textiles used to control noise in vehicles must provide airborne transmission reduction, damping and sound absorption. However, it is important to note that the use of acoustic textiles in vehicles is not only dependent on their acoustic properties but also on additional characteristics. Abrasion, stain, heat and UV resistance, aesthetics and texture, formability, recyclability, low weight, durability, antistatic finishing and tear strength are some of the characteristics that determine the choice of automotive textiles. Selection of a particular material is also determined by its ratio between performance and cost [4].

3.1 Seating Area

A large amount of vehicle interior sound absorption is provided by the seating area in modern vehicles [4]. The total sound absorption depends on the combination of

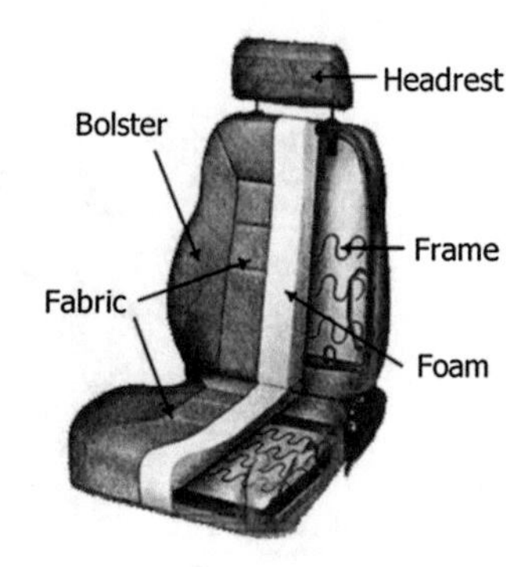

Fig. 5 Simplified diagram of the typical components of a car seat

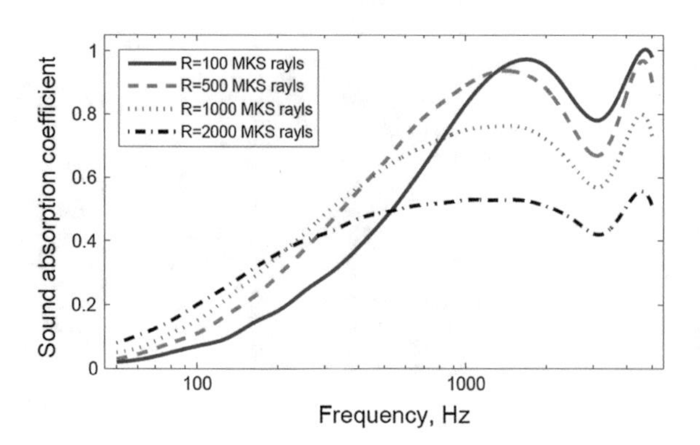

Fig. 6 Effect on the normal incidence sound absorption of the airflow resistance (R) of a nonwoven textile used as facing of a 50 mm thick porous foam

materials used to make the seat. Nylon, polyester, polypropylene and some natural fibres have been used as raw materials for seat covers. The use of woven and knitted textiles (such as tricot, double needle bar Raschel and circular knit) is common in American, European, and Asian seat upholstery manufacturers [4]. Figure 5 shows the typical components of a car seat.

Textiles are currently being used as replacement of polyurethane foam in the laminate of car seat covers, in particular using recycled fibres as cushion material. These include polyester fabrics, knitted structure, recycled wool and natural fibres such as coconut fibre, jute, sisal, kapok, some rubberized animal hairs, and three-dimensional nonwovens. Some of them also provide extra moisture permeability [31]. Double needle bar Raschel is widely used for both seat inserts and bolster areas.

In general, the textile seat cover acts as a porous facing to bulk foam, and depending on its flow resistance, it has the effect of increasing the sound absorption at low frequencies and reducing the sound absorption at higher frequencies as observed in Fig. 6. It has been noted that unless the percentage of the open area of the covering is 20 % or more, the reduction of sound absorption at high frequencies may be significant [32].

Because of its high cost, the use of leather as seat cover is more prevalent in luxury cars. Man-made leather and suede are also widely used by car manufacturers, and their manufacture requires textile base materials such as non-woven polyester microfibres [4].

3.2 Headliners

Basically, modern headliners are made of a combination of materials that include at least a rigid base structure to be attached to the metal panel (semi-rigid polyurethane foam, fibre-reinforced porous polymer, fibreglass, resinated shoddy fibres, cardboard), an intermediate soft layer (polyurethane foam, polyester nonwoven, polyolefin foam, polypropylene foam), and an interior decorative roof trim cover face (polyester nonwoven, knitted nylon/polyester, PVC foil) [4, 23]. The layers are bounded using hot melt adhesives, and then the structure is press molded to the required shape. These structures provide airborne noise isolation, structural rigidity and damping of the structural modes of the roof metal panel.

PET laminates used as decorative roof trim have been suggested for improving headliner sound absorption [31]. A PET nonwoven is laminated to an air permeable PET bonded by thermal activated bico/PET fibres.

3.3 Side Panels

Although door trims of lower- to middle-end cars are usually made of engineered plastics (PVC and polyolefin), textiles are commonly used in higher end vehicles [31]. Here, textiles require additional high formability to conform irregular shapes of door trims [4].

3.4 Carpets

Floor covering through carpeting, including tufted mats (made of nylon, PET, or polypropylene), represents a higher amount of textile acoustic surface used in an average modern car [4, 23].

Commonly, vehicle carpets are backed using viscoelastic polyurethane foam, resinated shoddy fibre/polyurethane foam, cotton fibre pads and an acoustic barrier made of ethylene, propylene and diene monomer rubber (EPDM) for noise reduction [4].

Needle-punched nonwoven (nylon, polypropylene, pigmented recycled PET fibre) is commonly used as surface material for interior carpets [15]. Natural fibre nonwoven floor covering systems for reducing automotive interior noises have also been described in literature [33]. Figure 7 schematically shows the main kinds of surface patterns for needle-punched carpet: plain, loop-like, and velour-like [31].

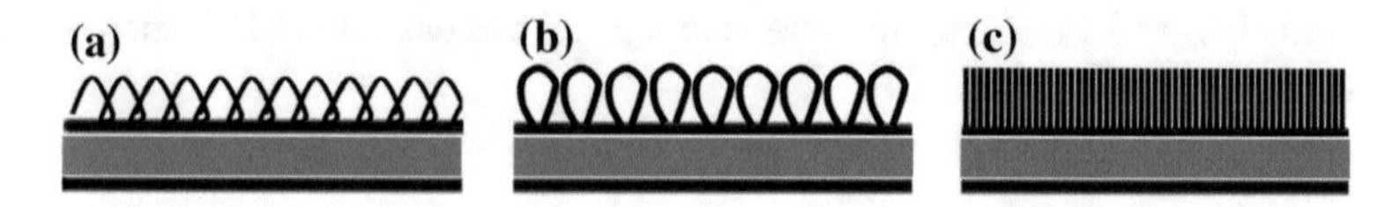

Fig. 7 Typical carpet surface patterns used for automotive interior lining: **a** plain, **b** loop-like, and **c** velour-like

The latter two patterns can be made by the combination of specific needles and needling patterns.

Low frequency sound produced by vehicle underbody fluctuating pressures is due to its large plan form area important for interior noise. Reduction of this noise is a real challenge for manufacturers since this kind of noise is difficult to insulate with floor carpeting since high surface mass is required to provide enough TL. Therefore, hard composite parts are becoming standard in the automotive industry. A classic example of an automotive composite panel is a fibre-reinforced plastic (FRP) composed of glass/carbon fibre and thermoplastic/thermoset resins, although some advanced composites have been recently tested [34, 35]. These specially designed materials are lighter and cheaper than steel and can be fabricated to conform irregular shapes using injection and compression molding as a fully integrated acoustic component.

3.5 Trunks

Textile molded parts used in trunks should provide a barrier of high TL to prevent airborne noise from the exhaust system within the car interior [4]. In addition, some vehicle audio systems include loudspeakers flush mounted into the package tray (and to other sidewalls also), where the trunk is used as a loudspeaker acoustic enclosure to get better low frequency response. In this case, sound absorption of the trunk reduces undesirable resonances, improving audio system quality.

Decorative yet functional polyester and polypropylene nonwoven are used for trunk lining attached to their front, rear and sidewalls. They require to have good wearing properties and to be easy to clean. Needled recycled fibres with elastomeric addition and spunbonded polypropylene can be sometimes used as a substrate for a foamed rubber material where trunk covering application is concerned [4].

3.6 Hood (Bonnet Liner)

Hoodliner (bonnet liner) is very important for reducing engine noise and structure-borne vibration. In the past, hoodliners were made of a bulk fibreglass mat covered with a protective PVC film. Higher standards in noise reduction and fuel

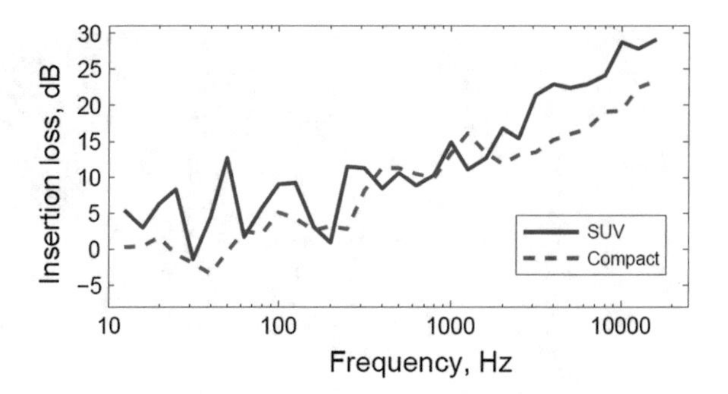

Fig. 8 Measured insertion loss provided by a hoodliner of an American SUV and a Japanese compact car

consumption have led to the development of composite multilayer structures for this purpose [36]. These structures are made of a combination of both conventional or recycled fibres and micro textiles designed to conform irregular shapes of the engine compartment of a specific vehicle. Textiles can be fluorocarbon treated to add water and oil repellent properties.

A common multilayer alternative used for reducing engine noise is made of a fibrous felt sandwiched between two layers of heat-resistant resin-impregnated nonwoven [31]. Figure 8 shows two examples of the insertion loss of a hoodliner. The insertion loss is the difference, in dB, between two sound pressure levels that are measured at the same point in space before and after the hoodliner is in place.

3.7 Other Applications

Package trays are also usually covered with needle-punched nonwoven textile mainly in polypropylene or polyester [4]. Special nonwoven wheel well inserts have also been developed to reduce tyre/road noise. This relatively new application of textiles has been found appropriate to replace PVC and EPDM in cars. Latex-coated needle-punched polyester and polypropylene are used in molded wheel arch liners to prevent the transmission of tyre/road noise and stone-impact noise to the passenger's cabin [37]. Better noise isolation and weight reduction have been observed when compared with old plastic liners, in particular on wet roads [4]. Pass-by noise may also be reduced with fibre wheel liners since they also provide sound absorption at the source of tyre/road noise.

In addition, fibre alignment relative to the incident sound causes the acoustic textiles to be anisotropic since airflow resistivity obviously changes for different angles. Consequently, sound absorption coefficient and TL usually differ for the case of grazing and normal incidence in lining applications. Fibre orientation also has a big influence on the tensile strength of a nonwoven material [38]. Figure 9a shows the fibre orientation of a cross section perpendicular to the face of a polyester nonwoven and Fig. 9b shows the pentalobe cross-sectional shape of a man-made fibre.

Fig. 9 Scanning electron microscope images of samples of fibrous sound absorbers: **a** fibre orientation of a cross section perpendicular to the face of a polyester nonwoven textile, **b** pentalobe cross-sectional shape of a man-made fibre (courtesy of J. Alba and R. del Rey, Polytechnic University of Valencia)

It seems that the friction between air and fibres can be increased for larger fibre surface area so that complex fibre cross sections may provide better sound absorption than circular cross-section fibres. Figure 9b shows an example of such a man-made fibre of a monofilament with longitudinally oriented grooves forming a pentalobe cross section. In Japan, a modified cross-section (trilobe) polyester fibre material was applied to the dash silencer, which produced improvement in sound absorption performance in a frequency range of 300–1000 Hz [39]. The dash silencer was attached to the dash panel dividing the engine compartment from the car interior to prevent the transmission of the engine noise to the passenger cabin.

The sound-absorbing properties of carbonized and activated nonwovens have been recently studied by some researchers. Textiles can be converted into active carbon products using pyrolysis, and they could be employed as high-performance sound absorbers [40]. Activated carbon fibre (ACF) fabrics have two levels of porous structures: macropores among fibres and yarns and micropores on the surface of ACF [41]. Fibre precursors for producing commercial ACF include rayon, acrylic, polyacrylonitrile (PAN) and novoloid [42]. So far, experiments on composites made of surface layers of ACF and fibres of cotton, ramie, and PP as base layers have shown high sound absorption coefficients in a wide range of frequencies [40]. These materials have potential to be used as high-performance sound-absorbing and noise isolators in automotive and other means of transport applications.

A very common design of a sound-isolating panel is a barrier typically made up of a perforated cover plate enclosing a core of a porous absorbing material and air gaps. Recycled materials can be used as core elements of these barriers. Recycled fibres have been used for many years as a source of raw material to produce acoustic textiles. Mechanically pulled waste clothing, known as shoddy, has been

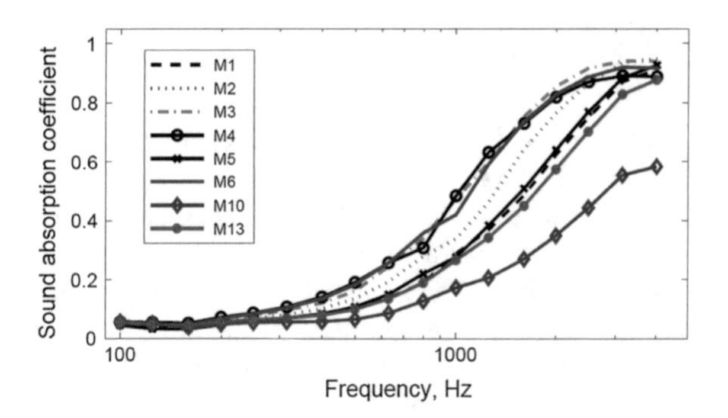

Fig. 10 Sound absorption coefficient at normal incidence of some recycled materials manufactured from textile waste [43]

Table 1 Properties of the recycled materials manufactured from textile waste

Sample	Multi-fibres	Binder	Density (kg/m^3)	Thickness (cm)	Airflow resistivity (kNs/m^4)
M1	80 % (75 % CO–25 % PES)	20 % PES	71	1.5	26.9
M2	80 % (75 % CO–25 % PES)	20 % PES	102	1.5	26.7
M3	80 % (75 % CO–25 % PES)	20 % PES	116	1.7	23.2
M4	74 % (75 % CO–25 % PES)	26 % PES	127	1.6	25.1
M5	74 % (75 % CO–25 % PES)	26 % PH	71	1.4	28.8
M6	74 % (75 % CO–25 % PES)	26 % PH	136	1.5	26.8
M10	80 % (75 % CO–25 % BP)	20 % PES	57	1.5	26.8
M13	80 % (75 % CO–25 % BP)	20 % PES	67	1.0	40.2

CO cotton, *PES* pure polyester, *BP* bico PET, *PH* phenolic resin [43]

used to produce porous nonwoven fabrics. Nonwovens made of recycled fibres that are used in laminated components incorporate thermoplastic fibres to allow molding to conform irregular automotive shapes. Figure 10 shows the values of sound absorption for recycled materials manufactured from textile waste [43]. Either hot melt fibres of recycled polyester or phenolic resin were used as binder. The properties of each of the manufactured materials are summarized in Table 1.

For vehicle applications that require high resistance to heat and fire (such as fire barriers), PANOX, Aramid and Kevlar are examples of raw fibres commonly used for this purpose [4]. Usually, they are combined with dense metal sheeting to provide higher values of TL.

4 Applications in Other Means of Transport

Although the largest amount of acoustic textile is employed in the automotive industry, other forms of transportation also use textiles to reduce noise and vibration. Some of these applications are discussed in this section.

4.1 Use in Aircraft

Interior noise levels in an aircraft cabin are often above 80 dB. This noise is produced by external sources (engines, propeller and turbulent boundary layer) and internal sources (air conditioning and hydraulic system, gearbox, landing gear, etc.) [28]. One of the main sources of interior noise in an aircraft is produced by engines. The nature of the noise depends upon the engine power and the mounting position of the engines (wing- and fuselage-mounting). Engine noise is transmitted into the cabin both airborne and structure-borne, where aircraft windows are usually an important noise transmission path.

Propeller noise is characterized by the presence of low frequency discrete tones at the fundamental blade passage frequency of the engines and their harmonics. For jet-powered aircrafts, the engine noise is predominantly broadband and the excitation depends on the mounting position of the engines. Interior noise in turbopropeller-driven aircraft is usually 5–10 dB higher than in a comparable turbofan-powered aircraft. Rotating imbalance forces within the engines can also cause noise in the interior cabin [44].

There is an important demand for reducing the interior noise in airplanes to enhance cabin comfort of both the crew and the passengers, in particular for long-haul flights. However, control of aircraft interior noise is difficult to achieve using passive techniques due to the restricted mass and volume constraints within an aircraft. This is particularly evident in turbopropeller-driven aircrafts, where the noise is usually dominant in the low frequency region (below 400 Hz) causing passenger discomfort.

The most common passive noise control technique in an aircraft is to improve the TL of the fuselage structure using multielement sidewalls. Basically, fibreglass sound-absorbing blankets are placed between the fuselage and the trim panels forming a double wall. Increasing the distance between the sidewall panels allows additional space for an air gap which increases the total noise reduction of the system [44]. Fibreglass blankets with a density of about 7 kg/m^3 contained in impervious bags are usually employed between the fuselage structure and interior trim panels as noise insulation [45].

Honeycomb panels are widely used in manufacturing parts of modern airplanes. A honeycomb panel is a lightweight sandwich panel with a honeycomb core [46]. This core can be manufactured with a variety of cell shapes, but the most commonly used shape is the hexagonal shape. Although costly, honeycombs are used because they have very high stiffness perpendicular to the plane and the highest shear stiffness- and strength-to-weight ratios of all available core materials [47].

Unitapes and woven fabrics have been used as composite face sheet materials of honeycomb panels. Common weaves include unidirectional fabrics, plain weave fabrics and satin weave fabrics. Plain weave appears to have the most stable construction, and its strength is uniform in both face directions [46, 47].

To add damping to the panel, a constrained viscoelastic layer is usually glued to a face of the panel [46]. This treatment is commonly used to dissipate vibration,

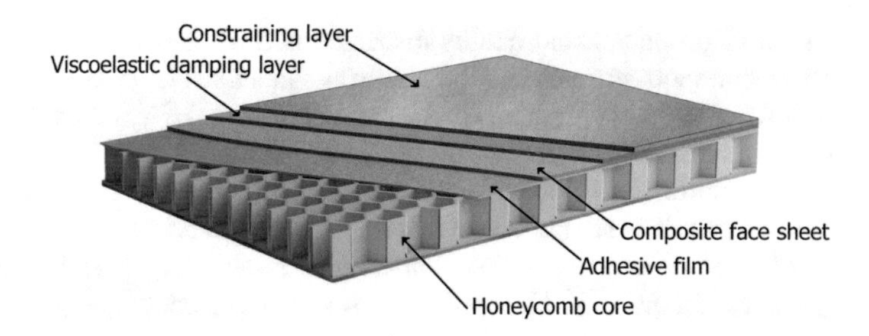

Fig. 11 Diagram of a typical lightweight honeycomb structural panel

thus reducing sound radiation from the structure [28]. Figure 11 shows a diagram of a typical honeycomb panel.

In the past, honeycomb sandwich cores were predominantly made of aluminium. Now these aerospace-grade honeycomb core structures are made of raw textile fibres such as Aramid, Kevlar and Nomex, which are coated with a heat-resistant phenolic resin [46]. These cores exhibit superior mass-to-stiffness ratios, higher strength, lower corrosion, fire protection, and better thermal, noise and vibration insulation. These honeycomb panels are mainly used in aircraft cabin floors, overhead bins, bulkheads, landing gear doors, engine nacelles and containment rings [28].

Carpet and seat upholstery in the cabin provides sound absorption, although treatment of the ceiling and bottom of the overhead bins is also used for this purpose. Carpets are generally made of woven loop pile wool with a polypropylene backing. Sound absorption material is often required in the neighbourhood of entry doors and galleys [48]. Furnishing of the cabin also helps to reduce the standing waves and resonances in the cabin cavity. It has been stated that a sound absorption coefficient greater than 0.2 is sufficient to suppress acoustic modes in a cabin [45]. Use of these textiles in aircrafts is limited by strict regulations related to wear, flammability, smoke and toxic gas emissions and antistatic properties.

4.2 Use in Trains

For most train interiors, the dominant airborne noise source comes from the wheel/rail interface, or rolling noise. Other sources of noise and vibration are from ancillary equipment, vibration of panels, leaks in body structure, a turbulent boundary layer that excites body panels and HVAC equipment [28].

Roughness of rails and wheels, especially corrugation in rails and out-of-round wheels, has been identified as a major cause of rail noise. The use of composite brake blocks rather than cast iron brake blocks will significantly improve the wheel running contact surface and reduce noise levels [49].

The concept of product sound quality discussed in Sect. 2 is also applicable to trains and it is an important concern for the railway industry. The train industry has been able to lower the sound pressure level in trains to a mean of less than 70 dB(A) (A-weighted decibel). A quantitative and qualitative study on the effects of noise sources on the interior noise characteristics in various types of trains has been recently reported [50]. The study showed that rolling, impact and curve squeal noises were caused by wheel–rail interaction. Impact noises have larger components at lower frequencies (below 1000 Hz) and curve squeal noises contain larger components at frequencies between 125 and 500 Hz in both underground and surface trains.

Pantographs in modern high-speed trains can produce significant aerodynamic noise reaching a noise level similar to rolling noise. Although noise from pantographs can be reduced by shielding and proper shape design, porous coatings can also provide significant noise reduction.

Honeycomb panels are very common in railway car body structures such as door panels, and in replacing conventional wooden interior floors, due to their low mass, resistance to moisture, and also to allow for electrical floor heating systems [46].

Sound absorption to reduce noise in train interiors is provided mainly by seat upholstery, loose coverings, carpets, curtains and bedding in sleeper compartments. Although woven fabrics made of wool and nylon are the most commonly used in train seats, the selection of optimal acoustic textiles is often restricted by fire requirements and vandalism protection [4]. In addition, polymer coating is usually applied to the surface of seat fabrics to confer impermeability to dust.

In some cases, barriers made up of perforated cover plates enclosing a core of sound-absorbing material may be used to control noise inside trains. An example of this type of barrier was described in Sect. 3.7. In this application, the perforated plate must face the train interior to provide sound absorption. These barriers can also be designed to achieve high values of sound TL and they have reduced flammability, durability and cleanability.

European and Japanese high-speed trains are replacing metal with composite structural materials to reduce weight, such as carbon fibre/epoxy, both in interiors and exteriors. Other fibres currently used in train insulation are Nomex and glass/phenolic lightweight composites [4]. Honeycomb panels to manufacture trains, subways and trams are widely used in modern transportation.

4.3 Use in Ships

Similar noise problems occur in passenger ships as in surface transportation vehicles and aircrafts. Major sources of noise in ships are the propulsion machinery and propellers. Due to the low damping of typical ship construction, airborne and structure- borne sound transmitted through the hull is difficult to reduce [28, 51].

There is a growing demand for acoustic comfort in ships. Subjective studies have indicated that acoustics is the most significant comfort criterion declared by passengers during a cruise [52]. Acoustics was also mentioned as the criterion to be improved, far ahead of other factors. The influence of acoustic discomfort on people is to produce sleep disturbance and irritation. The study showed that the most irritating noises for cruise passengers fall into three categories: (a) Squeaking, clattering, cracking, creaking noises, (b) Noise of engines, and (c) Ventilation and whistles.

A technique for noise and vibration reduction in ship cabins is the use of a floating floor. In some cases where cabins are located above extremely noisy rooms such as engine or auxiliary machinery rooms, floating floor may be the only alternative for reducing the noise levels in the cabin. Commonly, the floating floor consists of upper board and mineral wool, which is in turn laid on the steel deck plate. Mineral wool can be replaced for recycled carpet waste. Mathematical models to predict insertion loss of floating floors used in ship cabins have been reported by some authors [53].

The turbine and engine room is usually the noisiest place on a ship, and noise levels can be reduced using composite multilayer structures similar to the layers used in a car hoodliner. In addition, honeycomb panels made of raw fibres are used in ship hulls, bulkheads, fixed walls and other interior structures of ships to reduce weight and avoid corrosion [4].

Corrosion-resistance composites are also being used internally in ships as space dividers and doors, replacing metal to reduce weight.

Carpeting in passenger ships provides sound absorption in cabins to meet the demanding low-noise requirements in rest areas, in particular in luxury cruise ships, where high noise levels are expected to adversely impact sleep, comfort and speech intelligibility of both the crew and the passengers [51]. Furnishing fabrics used in ships such as Pyrovatex-treated cotton, Trevira CS and Fidion polyesters, comply with strict fire retardant standards [4].

4.4 Use in Spacecraft

A large number of materials used in spacecraft applications have been developed primarily to reduce noise and vibration exposure of humans, equipment and payloads. Use of any material in spacecraft needs to comply with very specific and accurate requirements. One of the important demands in aerospace applications is the reduction of weight and consequently, launching costs.

Intense acoustic noise and vibration are both unavoidable and undesirable by-products generated by the launching of a spacecraft. The noise generated during the firing of rocket engines manifests itself to the launch vehicle, sensitive spacecraft and the launch pad, in the form of airborne acoustics and structure-borne vibration [54].

The main sources of noise and vibration in spacecraft are from rocket engines, supersonic jet exhaust, pyroshocks, acoustic loads at launch, on-orbit jitter, and sometimes, planetary landing loads [54]. Due to the high intensity levels, spacecraft is subjected to the most hostile and extreme noise and vibration environment of any transportation system.

Composite textiles made of two or more fabrics bonded together using resins and adhesives have been developed for the spacecraft industry. Composites are designed to form an overall material whose properties are better than the sum of the properties of their individual components. For example, differences in the way the textile fibres are disposed may provide different strengths to a composite material. Carbon, glass and Aramid fibre reinforced materials are some of the composites that have been used for demanding spacecraft applications [3].

Spacecraft and payload structures are not the only things subjected to noise and vibration that could cause a mission to fail; other conditions are also quite important to consider, such as a hard vacuum, extreme temperatures, electromagnetic radiation and direct penetration by micrometeoroids and debris.

Multilayer textiles in the form of blankets have been used in a wide variety of aerospace applications, including launch vehicles, orbiters, satellites and the international space station. Some of the engineered fabrics on the market that have been used for these purposes are Kevlar, Beta Cloth, Orthofabric, Nomex and Nextel. Acoustic blankets have been used in the payload fairing of expendable launch vehicles to reduce the fairing's interior acoustics and subsequent vibration response of a spacecraft [54]. For example, a special acoustic blanket was developed and tested to reach NASA's goal of reducing the Titan IV acoustic environment to allowable levels for the Cassini spacecraft and Cassini's on-board electric power source, known as the Radioisotope Thermoelectric Generator [55]. Other high-temperature applications of these woven ceramic fibres and technical textiles for spacecraft are devised to protect liquid engine and pressurized gas lines from rocket plume.

On the other hand, smart textiles that are environmentally responsive were first developed by NASA to protect astronauts and instruments in the Apollo lunar landing program from the wide fluctuations of temperature in outer space [56]. These heat and flame-resistant synthetic fibres called polybenzimidazole (PBI) have an exceptionally high melting point and they are still used in the aerospace industry for making spacesuits and fire-blocking layers in airplanes seats [56, 57].

5 Conclusion

Although the transportation sector is currently the largest consumer of technical textiles, new developments and economic growth are expected to increase the development of more advanced acoustic textiles in the future.

Public concern on environmental impact of transportation is leading to reduced fuel consumption and the use of recycled materials. These are clearly related to the reduction of weight, extending durable years. Advances in multilayer composites are providing better performance, and producing lightweight materials. In this sense, the use of carbon fibre reinforced composites is very promising. Reuse of recovered materials and the use of environmentally friendly materials for vehicle manufacturing will be increased in the future. Natural fibres are being carefully examined as an option, and some benefits have been recently identified.

Although several researchers have established a number of theoretical and semi-empirical approaches in order to predict the acoustic characteristics of fibrous materials, further research on this topic is still required. More precise and reliable prediction models will allow us to improve the engineering of advanced textiles, consequently expanding their use to design better acoustic textiles.

The development of smart textiles will continue to be an important research subject. New textiles will be produced to incorporate electrical circuitry by using conductive threads knitted or woven into a fabric. Thus, the textile could be used as a sensor to provide information to digital signal processors, such as active interior noise control systems.

Therefore, the challenge that engineers will face is to create an acoustic material that supplies all of the aesthetic, technical, economical, ecological and security features that are required for use in modern means of transport.

References

1. Nayak R, Padhye R, Sinnappoo K, Arnold L, Behera BK (2013) Airbags. Text Prog 45:209–301
2. Mukhopadhyay SK, Partridge J (1999) Automotive textiles. Text Prog 29:1–125
3. Arenas JP, Crocker MJ (2010) Recent trends in sound absorbing materials. Sound Vib 44: 12–17
4. Fung W, Hardcastle M (2001) Textiles in automotive engineering. Woodhead Publishing Limited, Cambridge
5. Ghosh R (2014) Nonwoven fabric and the difference between bonded and needle punched non woven fabrics. IOSR J Polym Text Eng 1:31–33
6. Shoshani Y, Rosenhouse G (1992) Noise insulating blankets made of textile. Appl Acoust 35:129–138
7. Shoshani Y (1993) Studies of textile assemblies used for acoustic control. Tech Text Int 2:32–34
8. Ballagh KO (1996) Acoustical properties of wool. Appl Acoust 48:101–120
9. Lee Y, Joo C (2003) Sound absorption properties of recycled polyester fibrous assembly absorbers. AUTEX Res J 3:78–84
10. Moholkar VS, Warmoeskerken MMCG (2003) Acoustical characteristics of textile materials. Text Res J 73:827–837
11. Lou CW, Lin JH, Su KH (2005) Recycling polyester and polypropylene nonwoven selvages to produce functional sound absorption composite. Text Res J 75:390–394
12. Na YJ, Lancaster J, Casali J, Cho G (2007) Sound absorption coefficients of micro-fibre fabrics by reverberation room method. Text Res J 77:330–335
13. Liu Y, Hu H (2010) Sound absorption behaviour of knitted spacer fabrics. Text Res J 80:1949–1957
14. Soltani P, Zarrebini M (2012) The analysis of acoustical characteristics and sound absorption coefficient of woven fabrics. Text Res J 82:875–882
15. Shahani F, Soltani P, Zarrebini M (2014) The analysis of acoustic characteristics and sound absorption coefficient of needle punched nonwoven fabrics. J Eng Fibers Fabrics 9:84–92

16. Tascan M, Vaughn EA (2008) Effects of fibre denier, fibre cross-sectional shape and fabric density on acoustical behaviour of vertically lapped nonwoven fabrics. J Eng Fibres Fabrics 3:32–38
17. Delany ME, Bazley EN (1970) Acoustical properties of fibrous absorbent materials. Appl Acoust 3:105–116
18. Sides DJ, Attenborough K, Mulholland KA (1971) Application of a generalized acoustic propagation theory to fibrous absorbents. J Sound Vib 19:49–64
19. Shoshani Y, Yakubov Y (1999) A model for calculating the noise absorption capacity of nonwoven fiber webs. Text Res J 69:519–526
20. Garai M, Pompoli F (2005) A simple empirical model of polyester fibre materials for acoustical applications. Appl Acoust 66:1383–1398
21. Alba J, del Rey R, Ramis J, Arenas JP (2011) An inverse method to obtain porosity fibre diameter and density of fibrous sound absorbing materials. Arch Acoust 36:561–574
22. del Rey R, Alba J, Arenas JP, Sanchis V (2012) An empirical modelling of porous sound absorbing materials made of recycled foam. Appl Acoust 73:604–609
23. Shishoo R (ed) (2009) Textile advances in the automotive industry. Woodhead Publishing Limited, Cambridge
24. Crocker MJ (2007) Psychoacoustics and product sound quality. In: Crocker MJ (ed) Handbook of noise and vibration control. Wiley, New York, pp 805–828
25. Crocker MJ, Béchet C, Zhou R, Lingsong H (2008) Comparison between road test, loudspeaker and headphones evaluations of the sound quality of automobiles. In: Proceedings of 9th WSEAS international conference on acoustics & music: theory and applications, Bucharest, Romania, 24–26 June 2008
26. Arenas JP, Crocker MJ (2007) Use of enclosures. In: Crocker MJ (ed) Handbook of noise and vibration control. Wiley, New York, pp 685–695
27. Crocker MJ (2007) Introduction to transportation noise and vibration sources. In: Crocker MJ (ed) Handbook of noise and vibration control, Wiley, New York, pp 1013–1023.
28. Crocker MJ (2007) Introduction to interior transportation noise and vibration sources. In: Crocker MJ (ed) Handbook of noise and vibration control. Wiley, New York, pp 1149–1158
29. Sandberg U, Ejsmont JA (2002) Tyre/road noise reference book. Informex, Sweden
30. Bernhard RJ, Moeller M, Young S (2007) Automobile, bus, and truck interior noise and vibration prediction and control. In: Crocker MJ (ed) Handbook of noise and vibration control. Wiley, New York, pp 1159–1169
31. Matsuo T (2008) Automotive applications. In: Deopura BL, Alagirusamy R, Joshi M, Gupta M (eds) Polyesters and polyamides. Woodhead Publishing Limited, Cambridge, pp 525–541
32. Beranek LL (1971) Noise and vibration control. McGraw-Hill, New York
33. Parikh DV, Chen Y, Sun L (2006) Reducing automotive interior noise with natural fiber nonwoven floor covering systems. Text Res J 76:813–820
34. Gliścińska E, Michalak M, Krucińska I (2013) Sound absorption property of nonwoven based composites. AUTEX Res J 13:150–155
35. Jayamani E, Hamdan S (2013) Sound absorption coefficients of natural fibre reinforced composites. Adv Mat Res 701:53–58
36. Mirjalili SA, Mohammad-Shahi M (2012) Investigation on the acoustic characteristics of multi-layer nonwoven structures. Part 1—multi-layer nonwoven structures with the simple configuration. Fibres Text East Eur 20:73–77
37. Wyerman BR, Jay GC (2007) Tire noise reduction with fiber exterior wheel arch liners. SAE Paper 2007-01-2247
38. Maity S, Singha K (2012) Structure-property relationships of needle-punched nonwoven fabric. Front Sci 2:226–234
39. Watanabe K, Sugawara H, Koshikawa Y, Oku S et al (1996) Development of a high-performance dash silencer made of a novel shaped fiber sound-absorbing material. SAE Technical Paper 960192

40. Chen Y, Jiang N (2007) Carbonized and activated non-wovens as high-performance acoustic materials: Part I noise absorption. Text Res J 77:785–791
41. Chen JY, Jiang N (2014) Fabrication and characterization of carbonized and activated cotton nonwovens. J Ind Text 43:338–349
42. Jiang N, Chen JY, Parikh DV (2009) Acoustical evaluation of carbonized and activated cotton nonwovens. Bioresource Technol 100:6533–6536
43. del Rey R, Bertó L, Alba J, Arenas JP (2015) Acoustic characterization of recycled textile materials used as core elements in noise barriers. Noise Control Eng J 63:439–447
44. Hubbard HH (1995) Aeroacoustics of flight vehicles, theory and practice, volume I and II. Acoustical Society of America, New York
45. Wilby JF (2007) Aircraft cabin noise and vibration prediction and passive control. In: Crocker MJ (ed) Handbook of noise and vibration control. Wiley, New York, pp 1197–1206
46. Kumar S (2012) Sound transmission properties of honeycomb panels and double-walled structures. Doctoral thesis, The Royal Institute of Technology, Stockholm
47. Zhou R (2009) Sound transmission loss of composite sandwich panels. Ph.D. thesis, Auburn University, AL
48. Tate RB, Langhout EKO (1981) Aircraft noise control practices related to ground transport vehicles. SAE Paper 810853
49. Clausen U, Doll C, Franklin FJ, Franklin GV, Heinrichmeyer H, Kochsiek J, Rothengatter W, Sieber N (2012) Reducing railway noise pollution. European Parliament's Committee on Transport and Tourism, Brussels
50. Soeta Y, Shimokura R (2013) Survey of interior noise characteristics in various types of trains. Appl Acoust 74:1160–1166
51. Fischer R, Collier RD (2007) Noise prediction and prevention on ships. In: Crocker MJ (ed) Handbook of noise and vibration control. Wiley, New York, pp 1216–1232
52. Goujard B, Sakout A, Valeau V (2005) Acoustic comfort on board ships: an evaluation based on a questionnaire. Appl Acoust 66:1063–1073
53. Cha S-I, Chun H-H (2008) Insertion loss prediction of floating floors used in ship cabins. Appl Acoust 69:913–917
54. Arenas JP, Margasahayam RN (2006) Noise and vibration of spacecraft structures. Ingeniare Rev Chil Ing 14:251–264
55. Hughes WO, McNelis ME (2002) Recent advances in vibroacoustics. Sound Vib 36:20–27
56. Sen AK (2008) Coated textiles: principles and applications, 2nd edn. CRC Press, Boca Raton, FL
57. Messenger J, Wilson H (2003) Textiles technology. Heinemann, Oxford

Application of Acoustic Materials in Civil Engineering

Sujeeva Setunge and Nirdosha Gamage

Abstract Excessive noise energy over 60 dB is found to be disturbing and any noise over 75 dB can be considered as noisy. Exposure to an extreme noise can be a health risk and can cause significant hearing problem. Sound proofing is an important aspect for maintaining the health and well-being of the occupants of residential or commercial dwellers as well as to minimize disturbances to the neighbourhood. The acoustic retrofitting or acoustic treatments are commonly used for domestic, residential or populated buildings in order to absorb unwanted noise echoing and to confine sound energy inside the room. Noise walls and improved pavement materials are widely used to attenuate highway and freeway noise whilst aerated concrete and concrete walls are commonly used in buildings to prevent aeroplane noise from entering into residential buildings. Acoustic baffle systems and panels are commonly used in public areas and commercial buildings. Most of the Environmental Protection Authorities (EPA) around the world have regulations for residential noise, originated from vehicles, lawn mowers, electrical tools, domestic heating/coolers and radio or sound systems to ease neighbourhood noise. This chapter reviews various sound proofing applications used in Civil and Building Engineering as well as common materials used by the industry.

Keywords Noise barriers · Domestic noise attenuation · Acoustic panels · Aerated concrete for sound proofing

S. Setunge (✉) · N. Gamage
School of Civil Environmental and Chemical Engineering,
RMIT University, GPO 2476V, Melbourne, VIC, Australia
e-mail: sujeeva.setunge@rmit.edu.au

N. Gamage
e-mail: nirdosha.gamage@rmit.edu.au

R. Padhye and R. Nayak (eds.), *Acoustic Textiles*, Textile Science and Clothing Technology, DOI 10.1007/978-981-10-1476-5_8

1 Introduction

The use of acoustic materials in civil engineering, and buildings cannot be underestimated. Unwanted sounds originated from various sources such as vehicles, machinery, aircrafts or even loudspeakers are sometimes disturbing, intolerable and hazardous. The consequence of exposure to unwanted noise can interrupt the ability to hear clearly or continuous exposure to hazardous noise can trigger hearing problems. Various authorities have identified tolerance levels of noise, based on its intensity, time, place and the duration. Although, building regulations in the nineteenth century mainly concentrated on the structural stability and fire protection, building regulations in the twenty-first century have started to realize the importance of the sound proofing/insulation [1].

Currently, most of the Environmental Protection Authorities (EPA) around the world have regulations for residential noise, originated from vehicles, lawn mowers, electrical tools, domestic heating/coolers and radio or sound systems [2, 3]. A noise level of 60 dB is considered as 'noisy' (Fig. 1). Residents, who dwell close to airports can be exposed to high levels of noise. Also, loud music and sound systems can generate heavy noise within the own domestic environment. For example: Safe Work Australia has reported that 28–32 % of the Australian workforce are likely to work in an environment where they are exposed to non-trivial [$\geq$85 dB(A)] loud noise generated during the course of their work [4]. Continuous exposure to a noisy environment can trigger various health problems such as headache, fatigue, disturbed concentration, rise in blood pressure or cause hearing problems or heart diseases. Currently, there are some regulations available for managing residential noise, for restricting roadside/highway traffic noise management. Occupational health and safety instructions and regulations are available to protect workers from industrial noise [5].

Therefore, acoustic treatments are necessary to isolate and to control excessive unwanted noise. Emphasizing the noise attenuation actions at the design level is a good preparation to restrict the amount of noise entering a dwelling from neighbouring properties or streets or aeroplanes while minimizing the cost of construction. Thick walls and floors can help restrict noise movement and reflect or absorb excessive sound energy [6].

This chapter reviews various acoustic treatments used in Civil Engineering, exploring different materials commonly used for them.

1.1 Sound Insulation Standards

A need for sound insulation had been imminent for many years, but a rating of sound insulation for dwellings as an international standard was first introduced in 1968 (cited in [1]). European countries adopt 100–3150 Hz range as traditional acoustic frequency range for building requirements. However, Nordic countries that

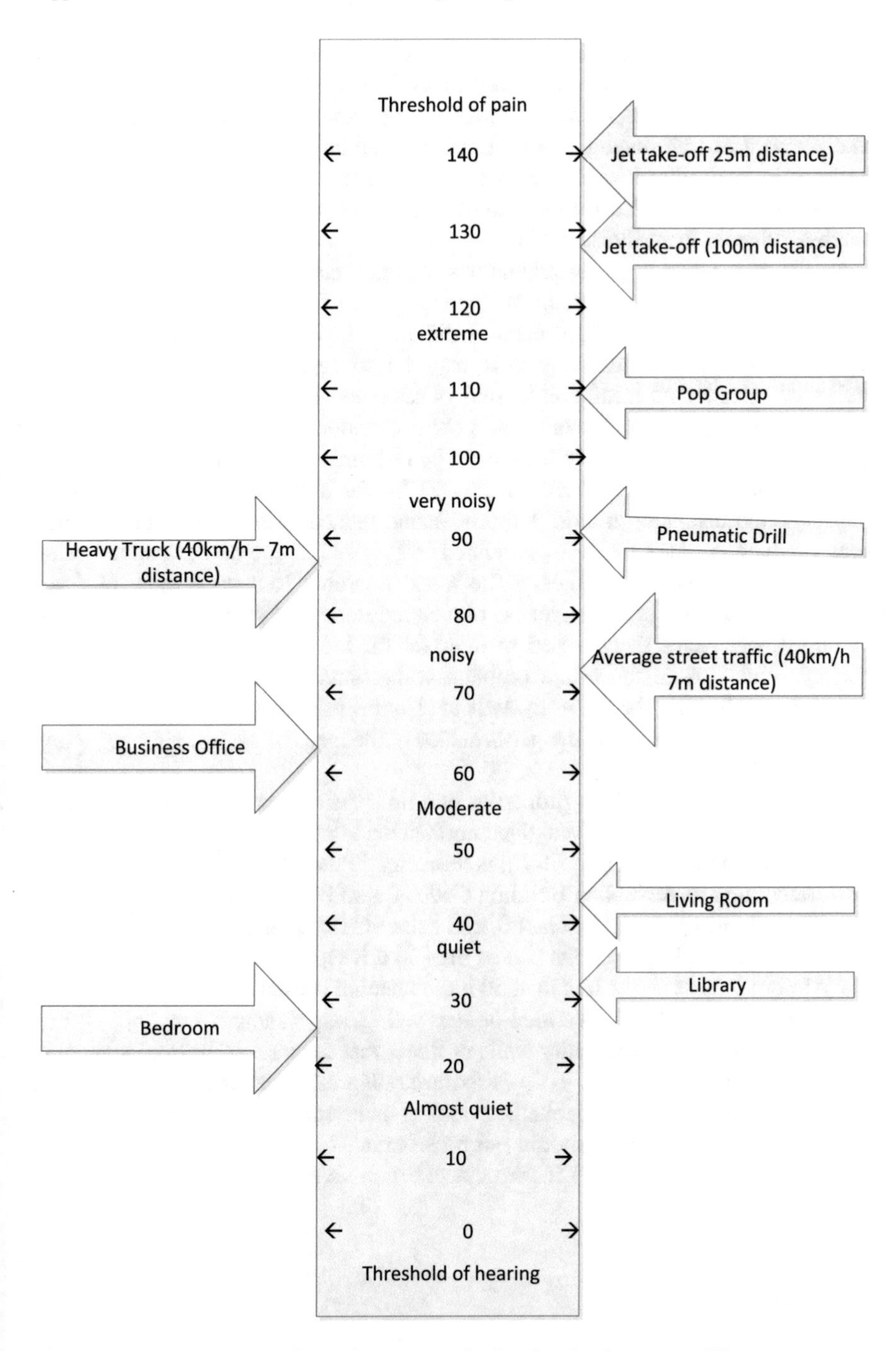

Fig. 1 An illustrative comparison of common noise level and noise sources [3]

use lightweight building material have introduced low frequency descriptors. The UK and Australia have taken a different approach that incorporate airborne traffic noise spectrum, C_{tr}-spectrum, which has a strong weight at low frequencies. The purpose of introducing C_{tr}-spectrum is to optimize sound insulation against traffic noise and other low frequency airborne noise.

The main purpose of sound insulation guidelines is to safeguard occupants and nearby residents from illness or loss of amenity as a result of transmitting severe noise [7]. Australian building regulations regulate transmitting exasperating noise between adjoining sole-occupancy units and from common spaces to sole-occupancy units and from parts of different classifications to sole-occupancy units. Under these regulations, it is required to separate floors, floor-wall or wall-wall using sound insulation barriers or retardants between sound sources such as plant rooms, lift shaft, stairway, public corridor and sole-occupancy units. Similar insulation is necessary for the walls. Determination of insulation rating for different material is determined by ISO 717.2 for airborne sound insulation or impact sound insulation rating. Airborne sound reduction index (R_w) or weighted sound reduction index ($R_w + C_{tr}$), defined in ISO 717.1, can be determined using results of laboratory measurements. The Weighted Sound Reduction Index (R_w) is a number used to rate the effectiveness of a soundproofing system or material. C_{tr} is an adjustment factor that is used to interpret for low frequency noise. Low frequency sounds make the prime problem with sound insulation. C_{tr} is always a negative number, so the $R_w + C_{tr}$ will always be less than the R_w value. These indexes are used to establish the performance of the material against airborne sound insulation [7–9].

Also, the sound insulation properties of a floor are compared against the impact sound insulation rating [10]. Weighted normalized impact sound pressure level with spectrum adaptation term ($L_{n,w} + C_I$) is measured in the laboratory using ISO 717.2 [10]. According to Australian Building Codes Board [7], domestic building wall or floor should satisfy class 2 or class 3, and class 9C for aged care buildings. A wall of class 2 or 3 building must not be less than 50 if it separates solo-occupancy units. R_w (Airborne) must not be less than 50 if the material is used to separate noise from a plant room, lift shaft, public area or stairwell. These standards must not be less than 45 for the aged care facility walls or floors that separate noise from spreading into or spreading out [7]. $R_w + C_{tr}$ (airborne) rating for walls or floor differs when the type of material in a habitable room is adjacent to a duct that transports wastewater. Then the value should not be less than 40 if the wall separates a duct and a habitable room, and 25 if the adjacent room is a kitchen or a non-habitable room.

1.1.1 Tests

AS ISO 354-2006 is an international standard to measure the sound absorption coefficient of the acoustical material used as wall or ceiling treatments or the equivalent sound absorption in an area of objects such as furniture [9]. Sound absorption coefficient (α_s), in one-third octave band (centre frequencies of 100–500 Hz inclusive), is a ratio of the equivalent sound absorption area of a test specimen divided by the area of the test specimen [9, 11]. If both sides of the specimen are exposed to the sound, then the sound absorption coefficient is the equivalent sound absorption area of the test specimen divided by the area of the two sides of the test specimen. The practical sound absorption coefficient (α_p) which is given in octave bands (centre of 125–4000 Hz inclusive) is calculated using the arithmetic mean of three corresponding α_s.

Noise reduction coefficient (NRC) is a standard coefficient, which defined in ASTM C 423-89, is the arithmetic average of the sound absorption coefficients measured in octave bands centred on 250–20,000 Hz inclusive, measured at four different frequencies: 250, 500, 1000 and 2000 Hz. NRC provides an indication of the absorbent noise performance of the material. Two materials with the same NRC perform differently in different applications because each may work best at a different frequency.

2 Various Applications of Acoustic Treatments in Civil Engineering

Acoustic treatments are used for a variety of buildings in various forms.

For example

1. Educational buildings, learning centres, common areas, auditoriums or lecture theatres.
2. Community areas such as churches, chapels, airports and travel hubs.
3. Entertainment rooms such as theatres, clubs and art galleries.
4. Commercial applications such as call centre cubical, meeting or conference rooms.
5. Residential settings such as home theatre rooms or houses that are near heavy noise sources like a freeway to control the internal or external noise level.
6. Residential buildings closer to airports or highways with excess noise.

One purpose of the acoustic retrofitting or acoustic treatments in these areas is to absorb unwanted noise echoing and to confine the sound inside the room, avoiding any disturbance to neighbours. Another purpose is to restrict the entrance of unwanted noise into the room and to avoid disturbances.

2.1 Acoustic Walls Panels and Floors

Noise wall is normally constructed using a lighter material with better sound insulation properties. Timber or porous concrete is normally used for

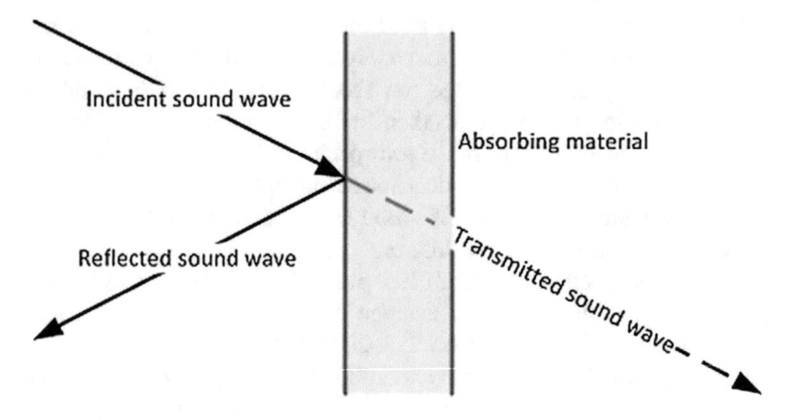

Fig. 2 Sound wave transmission through a material

highway/freeway noise attenuation walls while porous concrete, cellular walls or cavity walls are common for domestic wall applications. The material can block as well as absorb the majority of the incident sound waves to attenuate the sound to an acceptable level (Fig. 2).

2.1.1 Highway Noise Barriers

The level and the intensity of the road traffic noise depend on many factors such as the type of vehicles, speed, type of the road surface/pavement, vehicle age, topography as well as driver behaviour. Noise attenuation walls are commonly used adjacent to highways and freeways to attenuate airborne noise from traffic. These barriers can reduce the excessive loudness from traffic noise by as much as half irrespective of the noise wall material. Noise barriers either block the sound waves transfer through walls, from one side to another, e.g. between floors and ceilings, or attenuate the intensity of the sound by absorbing the noise (Fig. 3). Absorbing the sound by the material is the most efficient way of attenuating the sound transmitting through it.

However, the height of the barrier, length and the continuity of the barrier increase the attenuation by blocking noise travelling over it (Fig. 3). Installation of a barrier reflects some of the sound energy to the other side, some scatters over the ground coverings (grass etc.) and some energy goes over the barrier to a far away wider distances. Parts of the reflected energy will diminish due to the longer path that it must travel but remainder will be responsible for slight increased noise at the other side of the road. Field studies have shown that, if all the noise striking a noise barrier were reflected back to the other side of a highway, the increase would be theoretically limited to 3 dB [12]. Therefore, the height, length and the material of noise wall are designed after modelling the expected noise closer to it as well as far away or other side so that the residents in these areas are covered.

Fig. 3 Highway noise barrier that has both wood panels and light weight concrete panels

2.1.2 Highway Noise Mitigation Treatments

In addition to typical noise attenuation barriers, there are several other treatments available for individual dwellings and residential areas. When an individual house is required to be noise controlled due to various reasons, other treatments are introduced in lieu of or in conjunction with other noise control measures. These are

- specific facades of buildings
- double glazing to living areas and bedrooms and
- special road pavements near the dwelling to reduce noise level.

Buildings can achieve a reduced noise level 10 dB(A) below the external noise level if the windows are closed [13]. However, special facades treatments, double glazing of buildings are more expensive compared to common treatments, supply of alternative ventilation arrangements to such buildings.

Highway noise from structure-borne noise pitches is high due to heavy trucks travelling at high speed. The structure-borne sound of highways is usually reduced by the road pavement material such as open grade asphalt pavements.

2.1.3 Noise Attenuation in Buildings

Special walls, panels conjunction with rough surfaces are used for library buildings to prevent external airborne noise while noise absorbing materials such as carpets and wall panels are used inside to absorb internal noise. Structure-borne

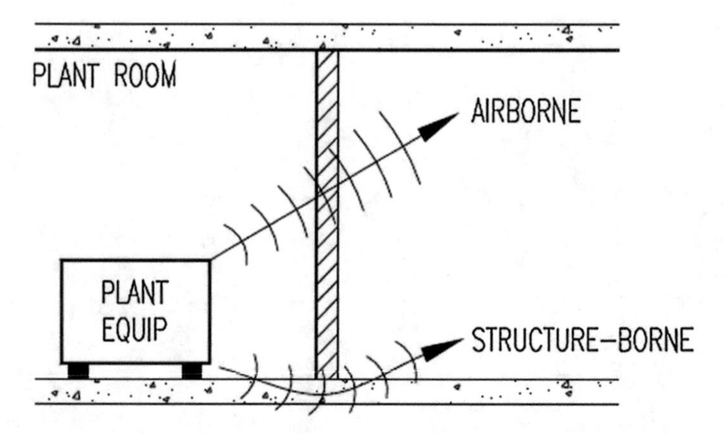

Fig. 4 Airborne and structure-borne noise [7]

(or impact-borne) noise, originates due to the vibration of a structure, e.g. vibration induced in the floor due to the vibration of the equipment (Fig. 4). Structure-borne noise is common inside or near machine rooms or closer to airports or near railway tracks.

Introducing a noise barrier between the noise source and the receiver can block transferring airborne noise to the receiver. Upper floor in a high-rise construction normally generates structure-borne noise that could disturb the floors below a level creating disturbing sounds. Attenuating such kind of noise cannot be undertaken by introducing a wall, but requires isolation of the noise and dampen it before transferring. Concrete floors are frequently cast-off between floors in the building near airports, commercial buildings or apartment buildings to prevent transferring noise from one floor to another.

Also, noise damping or insulation material is used between floors to dampen noise from spreading between floors. Alternatively, lightweight acoustic materials such as 'Hebel' or 'Hyssil', are high-performance materials that can be used for walls for attenuating and transferring noise from floor to floor, or from external to internal or the other way. Double walls or double glaze windows are common for such applications.

Soundproofing is necessary to restrict spreading of internal noises such as disco music or home theatre noise to other rooms or neighbouring houses. However, selecting the soundproofing materials for home theatre rooms is restricted due to the limited doors, windows and ceiling heights. Wall material and cavity walls are predominantly utilized for sound insulation. Decoupling ceilings or floor for sound insulation to prevent spreading the sound to an upper floor or lower level is limited. Building standards recommend providing insulations to connections of wall-wall or wall-ceiling to attenuate sound from travelling [7]. However, floor coverings such as carpets and rugs act efficiently to diffuse sound, and that can be upgraded further with an acoustic dampen underlay.

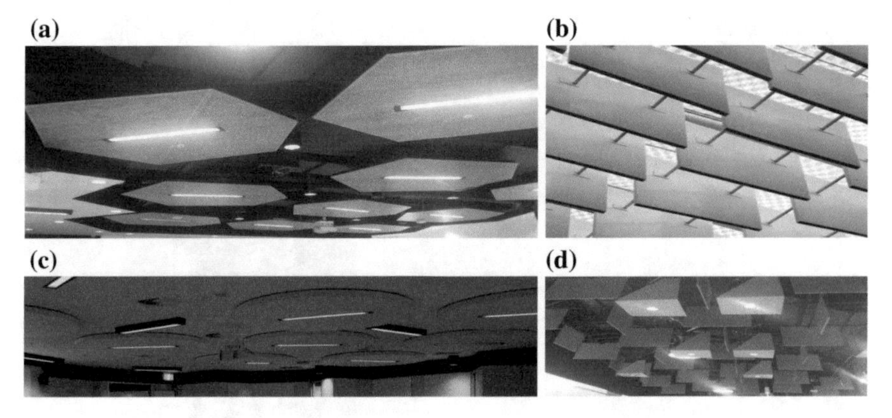

Fig. 5 Acoustic baffle systems inside lecture rooms (**a** and **c**) and public spaces. **a** Architecturally designed horizontally oriented (perforated material). **b** Vertically oriented baffles. **c** Architecturally designed horizontally oriented (foam material). **d** Horizontally, vertically and incline-oriented baffles

2.2 Acoustic Baffle Systems

Acoustic baffle systems are widely used in conference rooms, shopping mall lobbies, lecture theatres, warehouses or even car parks to reduce the internal noise level. Acoustic baffle systems are usually made of panels of sound absorbent material such as wood panel, cellular foam material, and they are used in large rooms and lobbies with ceilings of an adequate height and a large volume [14]. Baffle systems are designed with continuously changing the size and the shape of the baffles and the stance between them to allow for maximum sound attenuation. The foam type lighter materials (Fig. 5) are chosen and the panel thickness, panel edge shape is considered to provide an efficient sound attenuation while the colour is chosen to add to an enhanced visual image.

In addition to specific applications, most building materials such as brick or cement block walls or carpet floors used in civil engineering applications have significant sound absorption properties. However, specifically improved materials are currently available in the industry for efficient soundproofing application and some of them are deliberated in the next section.

3 Commonly Used Sound Insulation Materials in Civil Engineering

3.1 Acoustic Wood Panels and Wood Composites

Wood and wood panels are commonly used as noise attenuation barriers for freeway noise management in Australia. Various countries use different standards for

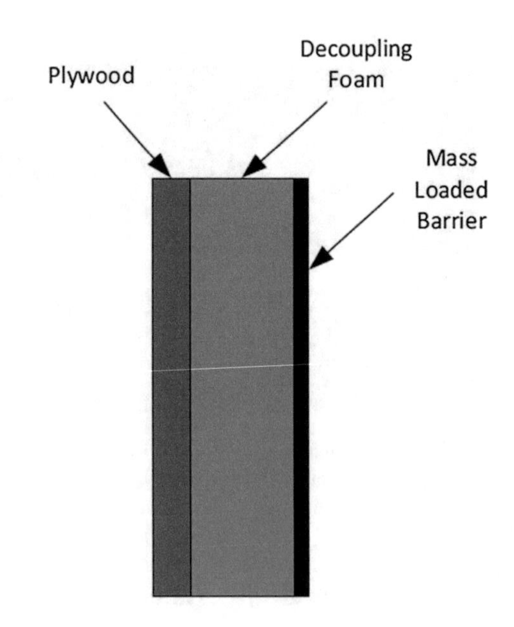

Fig. 6 Layout of plywood-foam test sample [18]

testing the suitability of the material. For example, EN 1793-2 is used by most European countries to evaluate the suitability of sound barrier material, based on measurements carried out in reverberant changers [15].

Timber-based noise attenuation barriers are commonly used in Australia to minimize disturbances to the nearby residence from freeway noise. Single-leaf, double-leaf or single-leaf with absorptive core are used for these noise barriers depending on the purpose and construction cost. The efficiency of the single-leaf noise barrier is largely dependent on the amount of air gaps in the timber panels through which sound waves transfer [15, 16]. The above investigation has demonstrated that the thickness of the panel or the mass density or the type of timber is not significant for the performance of a noise barrier but the percentage of air gaps in of the wood control its ability to prevent transfer of the sound.

Timber or composite timber materials are commonly used for noise attenuation treatments in various indoor applications such as machinery enclosures, intertenancy walls and bulkheads. Combining different timber elements in different forms has been investigated by many researchers in the past [17–19]. The composite material (Fig. 6) and the connection method used for composite action have also affected the efficiency of noise attenuation [14]. Plywood, foam composite with mass loaded barrier has improved the sound transmission loss properties of lightweight panels, similar to double-leaf walls. Different attachment methods: connection pins, spot glued and complete glue, were investigated in the research and transmission loss was measured. The results established that the decoupling of the mass loaded barrier from the panel has produced a significant increase in the sound transmission loss compared to simply increasing the surface density of a panel by the same amount. Moreover, by changing the connection method has significantly

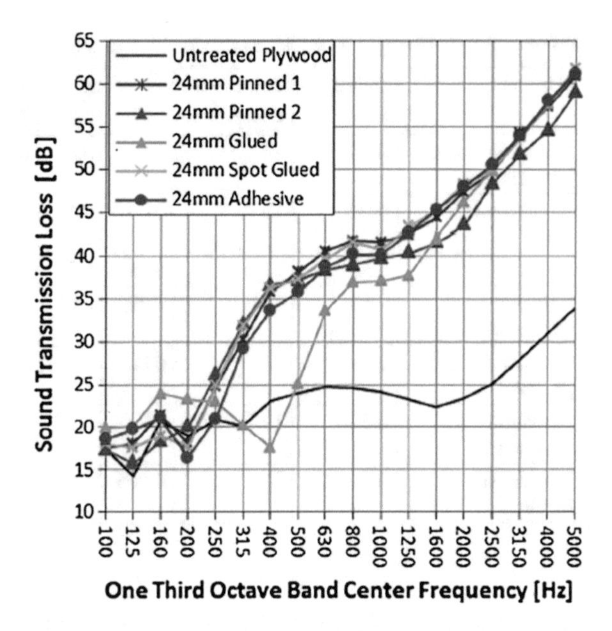

Fig. 7 Variation in the measured sound transmission loss due to the attachment [18]

reduced the noise transmission (Fig. 7). Complete attachment or stiffer attachment such as complete glue connection facilitated noise transmission and had produced lower transmission loss. Having air gaps in between panels or within the panel has pronounced the noise transmission lost for other materials too [15, 16, 20].

Particleboards are usually made using low-density wood and softwood waste combined with formaldehyde resin or isocyanate resin. Particleboards are heavily used for domestic building floor underlay and as floor panels for multi-story domestic construction. Low-density particleboard manufactured using low resin content has been investigated by many researchers for various applications, other than structural floors, such as for sound and thermal insulation properties purposes [21–25].

Table 1 compares the sound absorption coefficient of the various wood composite materials used in the building industry. The sound absorption coefficient of a material, α is defined as the ratio of the sound energy absorbed and the total energy impact [22]. Sound absorption of a material is dependent on the board density, thickness of the board as well as the intensity of the incident sound wave. The data shows that the low-density fibreboard and binderless particleboard have better sound absorption properties for high frequency sounds. This is due to the porosity of the low-density board, which provides functional drag, thereby reducing sound energy to heat. Whereas, commercial particleboard and plywood that have higher density compared to the general purpose boards have better sound resistance for low frequency sounds. The thickness of the low-density board can be adjusted to suit the intended application to absorb the intense sounds.

Table 1 Sound absorption coefficient of wood-based materials [22]

Material	Density (g/cm^3)	Thickness (mm)	Sound frequency (Hz)					References
			125	250	500	1000	2000	
Kenaf binderless particleboard	0.15	12	0.02	0.05	0.17	0.43	0.64	[22]
Wood board (pine)	0.52	19	0.09	0.1	0.12	0.08	0.08	[22, 24]
Plywood	0.55	12	0.25	0.14	0.07	0.04	0.1	[24]
Commercial particleboard	–	20	0.26	0.08	0.08	0.06	0.08	[23]
Low-density particleboard	0.3	30	0.06	0.15	0.37	0.65	0.52	[23]
Insulation fibreboard	0.22	12.7	0.04	0.06	0.14	0.38	0.69	[24]
Low-density fibreboard	0.2	12	0.006	0.02	0.08	0.35	0.71	[25]

3.2 Autoclaved Aerated Concrete (AAC)

Autoclaved aerated concrete (AAC) or Cellular lightweight concrete is a type of concrete that is lighter than average conventional concrete that in some cases will still provide similar strengths to lower grades of normal strength concrete. The basic idea behind the cellular lightweight concrete production is the addition of pre-formed stable foam into the concrete mixer that is already filled with slurry raw materials (such as cement, aggregates and sand) and water. The foam will impart foamed bubble that enables the concrete mixture to compact and expand. Therefore, cellular lightweight concrete may also be known as foamed concrete. The basic concept of manufacturing cellular lightweight concrete is to create a porous microstructure through entrainment of air in the concrete mix. It is done with addition of pre-formed bubbles or chemical admixtures which create a reaction to generate air bubbles [26].

These created pores are different to the pores created by hydrating cement paste during the early ages of concrete. The material has generated interest in the construction industry due to many advantages offered by the possible reduction of self-weight in concrete structures. AAC has been developed in Europe over a half century ago has been used for many civil engineering and building applications. Aerated concrete provides strong structural properties while reducing the density by 60–75 % compared to normal weight concrete.

In addition to the reduced self-weight, there are many other advantages of using the material that are better sound insulation properties, thermal insulation properties, fire resistance and easy handling [27]. High porosity has greater resistance to heat and sound transport [28]. Hebel wall panels and floor panels are a product of AAC that have been used mostly by European countries, Australia and New

Zealand for non-structural walls and residential floors. Hebel Acoustic Wall Systems (AWS) provide excellent barriers and insulator for residential and commercial application with achieving and R_w rating of 50 dB or better [29].

3.3 Acoustic Foam Products

Multi-porous polyester-based sound absorbing materials have been investigated by many researchers because of easy processing with pre-designed foam to achieve desired properties [30–32]. Noise absorbent behaviour of the multi-layer plastic microcapillary film (MCF) polymer has been investigated and its properties have been compared with two commonly used sound absorbent materials [33]. Acoustic performance of perforated MCF panels was compared with perforated galvanized steel plate and a porous material (Fig. 8). In measuring the perforation properties of the three samples, diameter for test samples was maintained at 96 mm. The thickness of the MCF samples was 3.5 mm for a the perforation rate of 1 %, perforation diameter of 1 mm, micro-capillaries diameter of 500 nm and a cavity depth of 50 mm were maintained. The perforated galvanized thickness of 10 mm was used when the perforation rate is 1.37 %, and perforation diameter of 1.2 mm and cavity depth 50 mm were used. The porous materials with sample thickness of 55 mm were used when pore size is 0.4 mm, and porosity is 20 %. Perforated galvanized steel performed better at 500 Hz frequency range whereas MCF performs twice better than galvanized steel at 800 Hz level. The porous material has comparatively low sound absorbent properties but performs consistently for a wider frequency range.

Polymer and plastic-based materials have been investigated for and commercially used in the sound reduction and noise absorbent purposes [34, 35]. These designed panel products have the ability to absorb sound efficiently at various frequency levels. These polymer-based sound absorbent panels typically perforate in two behaviours: to be used alone with a reflective surface to provide a narrow-band sound absorption and also used as facing of fibrous materials to provide sound absorption over a wider spectrum [36]. A perforated panel provides protection and the stability to the panel while the porous medium inside the material provides the sound absorption (Figs. 8 and 9).

In a separate investigation, the sound absorption property of highly porous, that is porosity of 86–90 %, titanium foam (Fig. 10) was investigated using standing pipe wave method [20]. This metallic foam material has shown a sound absorption coefficient over 0.6 in very high sound wave frequency 3150–6300 Hz. Having a low bulk density and stable properties, it may be possible to use the material in a variety of industries such as medicine, aerospace, electronics and theatres or transportation.

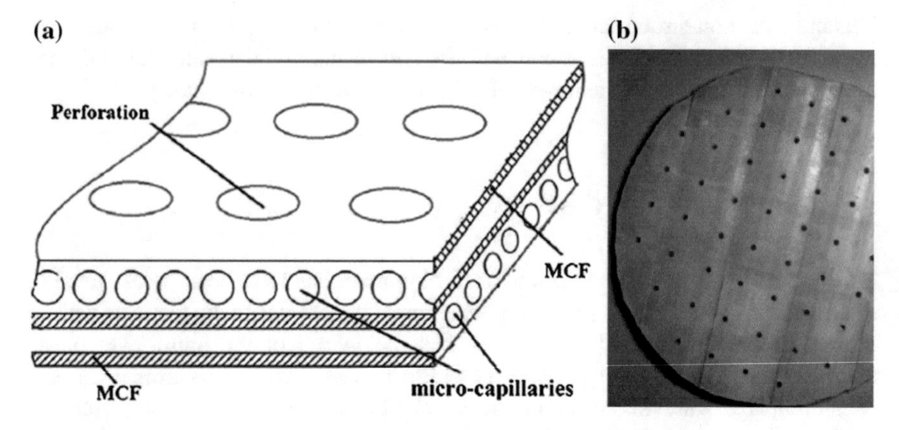

Fig. 8 **a** Cross section of MCF-based sound absorber; **b** photograph MCF [21]

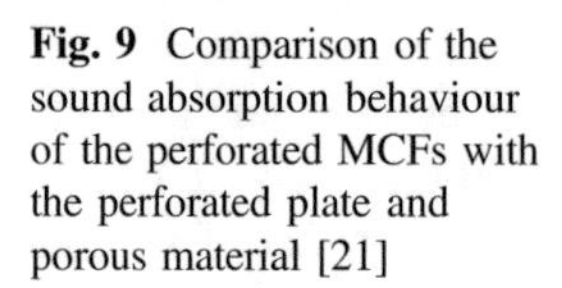

Fig. 9 Comparison of the sound absorption behaviour of the perforated MCFs with the perforated plate and porous material [21]

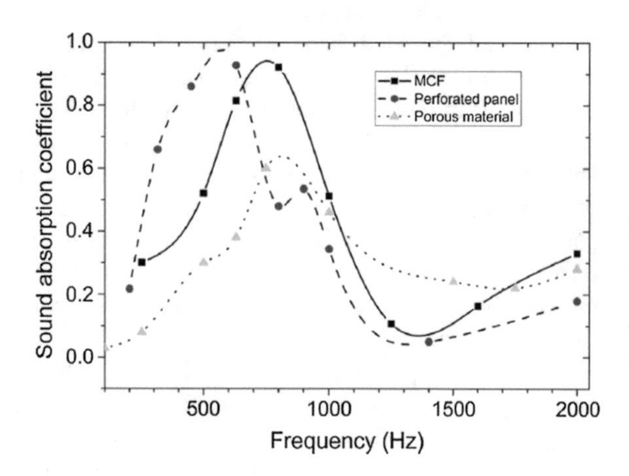

3.4 Woven Fabric and Fibre Retrofitting for Walls

Sound absorbing material coupled with a hard wall is widely used for sound insulation for closed spaces. Also, these materials increase the transmission loss properties of many applications such as in theatre rooms, machine rooms and various other functional buildings. Coconut coir fibre mats, recycled wool fibre mats, carpets and fabric, assemble to strong walls and floors have proven to reduce the noise transmission through the wall and reduce the reverberation by improving the sound absorption coefficient significantly [18, 38–40]. Highly porous and woven nature of these materials can absorb noisy sounds significantly (Fig. 11).

Fig. 10 Titanium foam samples for testing sound absorption coefficient [37]

Fig. 11 Fibres and woven mats for sound absorption. **a** Wool fibre [39]. **b** Coconut coir [41]. **c** Hard-wall retrofitter

3.5 Noise Absorbing Road Surfaces

The type and roughness of the road surfaces and the speed of the vehicle are main factors responsible for traffic noise generation. Rough and irregular surfaces contribute more to vibrate and emit noise compared to smooth surfaces. However, smooth surfaces sometimes contribute to traffic noise generation due to lack of pumping of entrapped air causing 'hissing sound'. Texture or fine irregularities within the road surface can facilitate the removal of entrapped air, without causing deformation or excess vibration of the tyre. Table 2 summarizes the sound escalation or mitigation properties of different road pavement materials compared to dense graded asphalt concrete pavements. Stone mastic asphalt and open grade asphalt provide significantly improved noise mitigation, hence they are commonly used in residential areas.

Table 2 Road surface noise corrections, relative to dense grade asphalt concrete [13]

Surface type	Noise level variation, dB(A)		
	Traffic noise	Individual vehicles pass-by noise	
		Cars	Trucks
14 mm chip seal	+4.0	+4.0	+4.0
Portland cement concrete: tyned and dragged	0 to +3.5	+1.0 to +3.5	−1.0 to +1.0
Cold overlay	+2.0	+2.0	+2.0
Portland cement concrete exposed aggregate	−0.05 to −3.0	1.01	−6.7
Stone mastic asphalt	−2.0 to −3.5	−2.2	−4.3
Open grade asphalt concrete	0 to −4.5	−0.2 to −4.2	−4.9

4 Other Benefits of Using Sound Insulation Material

Most of the sound insulation materials are good thermal insulators as well. Thermal conductivity is the property used to measure the thermal insulation of a material. A porous material and material with low thermal conductivity have better resistance to conduct heat, hence higher heat loss and act as a barrier between two sides. This is achieved by the air within the closed cellular structure resisting large amounts of heat being conducted through the material. This is quantified through the thermal conductivity that is a measure of how easily heat can transfer through an object [28].

The thermal conductivity of conventional AAC and conventional concrete are 0.11 and 0.38 Wm^{-1} K^{-1}, respectively. However, adding different additives or cement replacement into the conventional AAC mix changes the thermal conductivity due to the chemical compositions of the foreign material. For example, Fig. 12 compares the density and thermal properties of normal concrete and different AAC materials and shows that the thermal conductivity of AAC is significantly lower compared to normal concrete. Moreover, the thermal conductivity can be different although the density of the AAC is almost the same. However, the thermal conductivity of AAC types is still significantly lower compared to normal concrete.

5 Summary and Conclusions

Building regulations in the twentieth century have started to concentrate on the soundproofing and sound insulation inside buildings to maintain the health and well-being of the occupants. A noise level around 60 dB equivalent to a noise from a car, is considered as a noisy environment, and regular, continuous exposure to a noisy environment or frequent sound can lead to hearing problems of occupants.

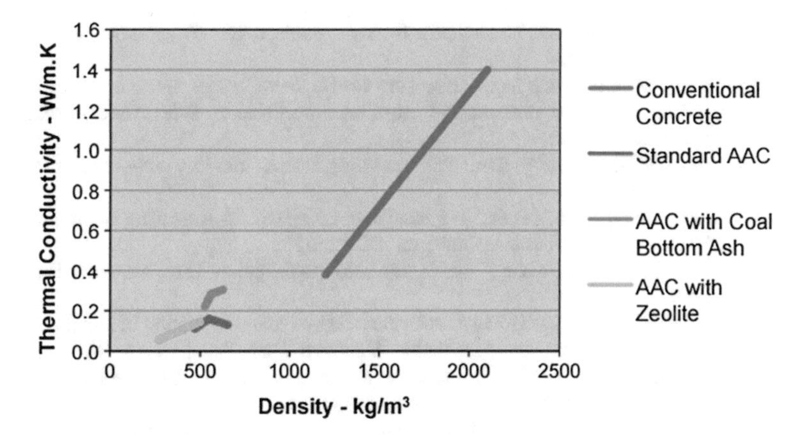

Fig. 12 Thermal conductivity and density among different AAC and concrete [28]

Noise is classified as airborne noise, which is the noise from a direct source of travelling through air and structure-borne noise that is initiated by vibration of structures. An airborne noise such as traffic noise can be attenuated by noise barriers, while structure-borne noise can be treated by modifying the structure or surface material such as road pavements.

There are two main standards that define acceptable levels of noise: European standards define the range of 100–3150 Hz for buildings. ISO 717.1 defines an acoustic sound reduction index R_w to be less than 50 in buildings.

The acoustic retrofitting or acoustic treatments are used for domestic, residential or populated buildings in order to absorb unwanted noise echoing and to confine the sound inside the room, avoiding any disturbance to neighbours.

Noise walls are commonly used to block freeway noise from entering into residential areas and various retrofitting treatments such as baffle systems, carpets and curtains are used to absorb noise and attenuate indoor noise.

The common feature of the most noise attenuation material is the low density and high porosity that can absorb high-intensity sound waves while minimizing the reverberation and the sound intensity. Common materials adopted are wood-based composites, aerated concrete, multi-layer plastic micro capillary films (MCF) and fibre and woven mats. Most sound insulation materials have good thermal insulation properties as well.

References

1. Rasmussen B, Rindel JH (2005) Concepts for evaluation of sound insulation of dwellings—from chaos to consensus. In: Forum Acusticum 2005. OPAKFI © 2005, Budapest, Hungary
2. EPA-VIC (2015) Noise legislation. Available from:http://www.epa.vic.gov.au/about-us/legislation/noise-legislation#noiseregs

3. EPA-NSW (1999) Environmental criteria for road traffic noise. Environmental Protection Authority, NSW, Australia
4. Australia SW (2009) National hazard exposure worker surveillance—noise exposure and the provision of noise control measures in Australian workplaces. Safe Work Australia for Commonwealth of Australia
5. Australia SW (2009) Managing noise and preventing hearing loss at work—code of Practice. Safe Work Australia
6. Board AABC (2004) Sound insulation guideline document, p 52. Australian Government and States and Territories of Australia, Canberra, ACT 2601
7. EPA-NSW (1999) Environmental criteria for road traffic noise. Environmental Protection Authority, NSW, Australia
8. Standard A (2004) AS/NZS ISO 717.1: acoustic rating of sound insulation in buildings and of building elements—airborne sound insulation. Standards Australian, Sydney, NSW, Australia
9. Standard A (2006) AS/ISO 354: acoustic-measurement of sound absorption in a reverberation room. Standards Australia, Sydney NSW 2001, Australia
10. Standard A (2004) AS ISO 717.2-2004: acoustics—rating of sound insulation in buildings and of building elements—impact sound insulation. Standards Australia, Sydney, NSW, Australia
11. Li A (2011) Measurements of sound absorption timbercrete, p 10. Timbercrete Pty Ltd., Bilpin, NSW, Australia
12. FHWA (2016) Highway traffic noise. Office of Planning, Environment, & Realty (HEP) [cited 22 Jan 2016]
13. Isles JACS (2016) RTA environmental noise management manual [cited 15 Jan 2016], Issue 1. Available from:http://www.rms.nsw.gov.au/documents/about/environment/environmental-noise-management-manual.pdf
14. WAVE-Baffles (2011) Wave design solution [cited 03 Aug 2015]. Available from:http://www.wave-akustik.de/WAVE/EN/wave_baffles_en.html
15. Watts G, Morgan P (2007) Measurement of airborne sound insulation of timber noise barriers: comparison of in situ method CEN/TS 1793-5 with laboratory method EN 1793-2. Applied Acoustics 68(4):421–436 (2006)
16. Watts G (1999) Effects of sound leakage through noise barriers on screening performance. In: Sixth international congress of sound Vibrat. Copenhagen, Denmark
17. Davy JL (2009) Predicting the sound insulation of walls. Build Acoust 16(1):20
18. Wareing RR, Davy JL, Pearse JR (2015) Predicting the sound insulation of plywood panels when treated with decoupled mass loaded barriers. Appl Acoust 91:9
19. Davy JL (2012) Sound transmission of cavity walls due to structure borne transmission via point and line connections. J Acoust Soc Am 132(2):8
20. Liu PS, Qing HB, Hou HL (2015) Primary investigation on sound absorption performance of highly porous titanium foams. Mater Des
21. Xu Z et al (2015) Design and acoustical performance investigation of sound absorption structure based on plastic micro-capillary films. Appl Acoust 89:7
22. Xu J et al (2004) Manufacture and properties of low-density binderless particleboard from kenaf core. J Wood Sci 50(1):62–67
23. Kawai S et al (1988) Thermal, sound, and fire resistance performance of low-density particleboard. Mokuzai Gakkaishi (cited in Xu et al 2004) 34:8
24. Kawai Y et al (1997) mokusituzairyowugaku (in Japanese). Towuyowu Syowutenn (cited in Xu et al 2004) 289:5
25. Kawasaki T, Zhang M, Kawai S (1998) Manufacture and properties of ultra-low-density fiberboard. J Wood Sci (cited in Xu et al 2004) 44:7
26. Lee HY (2010) Compressive strength and drying shrinkage of HySSIL cellular lightweight concrete under different exposure condition. School of Civil, Chemical and Environmental Engineering, RMIT University
27. Memari AM, Chusid MT (2003) Introduction to architectural aspects and developments in research on structural performance of autoclaved aerated concrete (AAC) products. In: Proceedings of architectural engineering 2003: building integration solutions, pp 1–5

28. Gates A, Marriage G Autoclave Aerated Concrete as a suitable alternative to conventional concrete in a seismically active location
29. Hebel C (2005) Acoustic wall system—designing for future living. http://www.peachester.com.au/hebel-resources/hebel-design-guides/old/wAWS_Design_guide_ef1b.pdf
30. Ahmad S (2003) Reinforcement corrosion in concrete structures, its monitoring and service life prediction—a review. Cement Concr Compos 25(4):459–471
31. Chen D, Li J, Ren J (2010) Study on sound absorption property of ramie fiber reinforced poly (l-lactic acid) composites: morphology and properties. Compos A Appl Sci Manuf 41(8): 1012–1018
32. Yin Y et al (2012) An investigation on polymer modulus test using laser-based finite element method. J Acoust Soc Am 131:3373
33. Xu Z et al (2015) Design and acoustical performance investigation of sound absorption structure based on plastic micro-capillary films. Appl Acoust 89:7
34. Falke P et al (2001) Production of sound-damping and energy-absorbing polyurethane foams. Google Patents
35. Megasorber (2015) Acoustic polyester panel with fireproof sound-absorbing facing [cited 29 July 2015]. Available from:http://www.megasorber.com/soundproofing-products/sound-absorbers/fireproof-faced-acoustic-polyester-wool.html
36. Hong Z et al (2007) A novel composite sound absorber with recycled rubber particles. J Sound Vib 304(1):400–406
37. Chen B, Liu J (2008) Experimental application of mineral admixtures in lightweight concrete with high strength and workability. Constr Build Mater 22(6):1108–1113
38. Nor MJM, Jamaludin N, Tamiri FM (2004) A preliminary study of sound absorption using multi-layer coconut coir fibers. Electron J Tech Acoust 3:1–8
39. Patnaik A et al (2015) Thermal and sound insulation materials from waste wool and recycled polyester fibers and their biodegradation studies. Energy Build 92:161–169
40. Soltani P, Zerrebini M (2012) The analysis of acoustical characteristics and sound absorption coefficient of woven fabrics. Text Res J 82(9):875–882
41. Ismail HKM, Fathi MS, Manaf NB Study of lightweight concrete behaviour. Universiti Teknologi Malaysia

Design of Acoustic Textiles: Environmental Challenges and Opportunities for Future Direction

Nazire Deniz Yilmaz

Abstract Textiles have begun to find use in acoustical applications for the past few decades. In order to consolidate the position of textiles as noise control materials, their performance characteristics should be further enhanced to a level comparable to conventional acoustic materials, the most common ones of which are glass fibre mats and polyurethane foams. The same recent history has also witnessed substantial growth in public concern related to environmental effects of industrial progress. As much as this situation imposes a burden on textile producers, it also opens a new battlefield against acoustic materials, where textiles can prevail. Conventional acoustic materials, mineral wool or polyurethane foam among others are under line of fire due to their adverse effects on the ecosystem as well as on human health. This situation can offer business advantage to textile producers, provided that the damage inflicted on the environment throughout the whole life cycle of the textile product is minimized and the functional properties are improved. This can be realized by implementing a sound design stage based on functional and environmental requirements related to acoustic textiles.

Keywords Acoustic textiles · Biodegradable · Design · Environmentally friendly · Life cycle assessment · Recyclable · Sustainability

1 Introduction

Noise pollution which was once considered as irritating but harmless, is now receiving more attention in the general community [1]. Noise is a major issue in various environments including dwellings [2], work offices [3] and vehicles [4]. Now it is already known that noise is a primary health problem showing undesired physiological effects such as stress-related illnesses, high blood pressure, sleep disruption and noise induced hearing loss, among other countless health problems

N.D. Yilmaz (✉)
Department of Textile Engineering, Pamukkale University, Denizli, Turkey
e-mail: ndyilmaz@pau.edu.tr

R. Padhye and R. Nayak (eds.), *Acoustic Textiles*, Textile Science and Clothing Technology, DOI 10.1007/978-981-10-1476-5_9

as stated by Environmental Protection Agency. Based on the fact, studies have revealed that people who are exposed to noise constantly at elevated levels are at risk of various illnesses [5]; accordingly, workers whose working conditions necessitate exposure to high levels of sound receive extra payment in Germany [6].

In addition to its effects on health, noise also leads to productivity losses in various work environments [1]. Even sound, that is not high level, such as speech and phone bell ringing present in regular work offices, results in disturbance of worker and reduces the work quality. This is especially the fact if the work involves tasks that are cognitive demanding, requiring high level of concentration or based on verbal communication [7]. Thus, exposure to unwanted sound causes the workers to take extra breaks, work overtime and overexert. Furthermore, it deteriorates the well-being of the workers by increasing their general level of stress symptoms as well as by impeding their job satisfaction. According to a survey conducted by Haapakanagas et al. [3], noise was found to be the major source of disturbance for indoor office environment surpassing problems related to thermal conditions and air quality.

Noise is also a major problem in transportation. In an automotive example, where noise control is a prime issue, the unwanted sound which is present in a vehicle's passenger compartment impedes the intelligibility of speech, leads to driver fatigue, and risks the occupant safety [8]. Additionally, the interior noise is a general complaint of car owners. Hence, the noise in a vehicle not only impedes the safety, but it also damages the quality perception of the vehicle [9]. Furthermore, noise control becomes even more important as the electronic communication and entertainment devices are increasingly integrated into vehicles [4].

Whereas noise impedes our health, productivity and comfort, human beings are not the only organisms that are affected by noise. Research has shown that marine mammals, bird species [10], plants and insects suffer from fatal effects of noise which may cause unwanted long-term effects on biodiversity and ecosystem structure [11].

The unwanted sound can be reduced by noise control methods which can be classified in three major groups: modifying sources of noise vibration, using barriers to disconnect the path of sound propagation and finally attenuate sound by use of sound absorbers. Sound absorption materials are porous materials wherein the sound energy is dissipated due to viscous and thermal losses that take place in tortuous pore channels. There also are three types of porous materials, namely cellular materials (foams), granular materials (woodchip panels, porous concrete, pervious road surfaces) and fibrous materials. Acoustic textiles which fall under the category of fibrous materials show promise for noise control applications [12, 13].

Prevention of noise pollution using acoustic textiles should not entail damaging the environment by depleting natural resources, releasing greenhouse gases, emitting hazardous substances into air, water or soil systems. Thus, the potential ecological impact should be taken into consideration during the design stage of acoustic textiles together with other functional performance characteristics. During this process, whole life cycle of the acoustic textiles should be analyzed, from material extraction to disposal or reuse after useful service life.

This chapter investigates acoustic textiles in terms of their environmental impact. The next section deals with the design stage referring to life cycle assessment methods. The second section examines the effects of manufacturing stage of the acoustic textiles. This section also reviews different materials used for sound absorption and ecolabels. The third and the fourth sections consider recycling issues and end-of-life concerns of acoustic textiles respectively. Finally, the fifth section concludes the chapter.

2 Design Challenges

During the design stage of an acoustic textile, the first major step is to determine the noise source. For example; for a road vehicle's passenger department, the engine, the wind and the road contact act as the major noise sources. This makes the hood insulation, the headliner, and the floor coverings as critical elements in terms of noise control [9]. The most important parameter other than the noise source is the audible frequency range of interest. In the case for passenger compartments of vehicles, medium frequency range (1200–4000 Hz) becomes critical [14]. Thus, the designed acoustic material should incorporate characteristics to allow effective sound attenuation in the targeted frequency range. Also, the sine qua non properties mandated by the application where the acoustic textiles should be used, such as moldability [15] and nonflammability [16] and durability in high temperature, high speed turbulent flow, [17], high humidity [16] and other hostile environments of interest should be taken into consideration. To give a civil engineering example, depending on the building class and purpose of use of the building, interior acoustic panels should meet stringent requirements such as that of A2 (noncombustible) or B (combustible-very limited contribution to fire) construction products according to EN 13501-1 the European classification for the reaction to fire behaviour of building products [16, 18]. Most of the time, acoustic materials, and acoustic textiles, are expected to serve multiple tasks, not only noise reduction. Thus the designed material should also perform well in thermal insulation, comfort or contribution to passive safety systems based on the other functions desired [9]. Other desirable features of acoustic materials in terms of ecology and economy may be given as recyclability, light weight and of course cost effectiveness [15, 4]. Table 1 summarizes this initial design stage of acoustical materials.

Once the material selection that caters for the mentioned requirements has been conducted, comes the turn to devise the material mix to provide the desired noise control effect.

Here, it is, of course, of paramount importance to possess sound knowledge pertaining to acoustical wave propagation through materials of interest in order to design noise control elements to serve in various fields [4]. To understand the acoustical properties of sound absorbers, various mathematical models have been developed. Majority of these models can be classified in two categories: empirical phenomenological and theoretical microstructural models. Moreover, transfer

Table 1 Steps in the initial design stage of acoustical materials

Step #	Explanation
1	Determination of sound source
2	Assessment of the frequency range of the unwanted sound
3	Designation of sine qua non properties mandated by the end use
4	Identifying the required characteristics to be able to perform functions other than noise control
5	Specifying desirable features in terms of ecology and economy

matrix modelling is a very helpful tool that is frequently utilized for porous materials with single or multiple layers [19].

A mathematical model that can predict the noise control capability using parameters such as material thickness, porosity, fibre content and other intrinsic characteristics is very useful for designing acoustic textiles. However, developing a theoretical model to account for the sound waves travelling in the pores of porous materials incorporates some difficulties. The solutions of the obtained equations ought to be subjected to boundary conditions ensuring continuity between neighbouring air and material phases [4]. The model also has to involve dynamics of fluids, acoustics and statistics [20]. Moreover, the huge number of parameters makes it nearly impossible to obtain agreement between theoretical prediction and experimental data [4].

Due to the inherent difficulties of the theoretical models [21], a good number of empirical models have been developed to predict the sound absorptive capacity of porous materials taking a macroscopic approach. The drawback of these models is their limited usefulness during the design phase as they may not include the microscopic parameters such as fibre size and fibre orientation [19]. However, there are also empirical models which study microscopic parameters as well as macroscopic parameters [13, 22, 23, 24]. The parameters which determine noise control capacity of acoustic textiles are listed in Table 2 [4, 25].

Design is a challenging task considering the fact that varying material types together with acoustic textiles can be used simultaneously in different shapes, thicknesses, sequences, perforation and grooves properties. For example, core materials (medium density fibre board (mdf), cement, particleboard, flake board, block board, forex), surface material (veneer, melamine, paint), besides specially developed materials, are used for building interior noise control elements [16]. Figure 1 presents a design lay out where effect of different layer sequences on sound absorption capacity was investigated.

As the sound absorbers are composed of multiple layers of different materials most of the times, modelling the acoustical behaviour may be tiresome. For this, the aforementioned transfer matrix modelling presents a very powerful technique during design stage that can be used for multiple—as well as single-layer absorbers [19].

A massive matrix of parameters is generated when the variations in material, processing and assembling parameters are considered. While conducting the

Table 2 Parameters affecting sound absorption performance of acoustic textiles [4, 25]

Category	Subcategory	Parameter
Material parameters	Macroscopic physical parameters	Air flow resistance and resistivity
		Impedance
		Thickness
		Density
		Porosity
		Tortuosity
		Fibre orientation distribution
		Composition
		Air gap
	Fibre parameters	Fibre type
		Fibre size
		Fibre shape
Process parameters	Production parameters	Web forming technique
		Web bonding technique
	Treatment parameters	Chemical treatment
		Physical treatment
Other parameters	Medium	Medium density
		Medium shear viscosity
		Medium heat conduction coefficient
	Sound	Frequency
		Angle of incidence

selection, the potential environmental impact of each variation should be conceived. In relation to that, the environmental assessment should be carried out beforehand in the design stage, not at the end of service life. To designate the environmental sustainability of a designed product, life cycle assessment (LCA) provides a useful tool.

2.1 *Life Cycle Assessment (LCA)*

To ensure the environmental-friendliness of a material, including acoustic textiles, contrary to the general opinion, neither biodegradability nor recyclability is sufficient. The ecological impact should be determined by considering a high number of criteria through the whole life cycle, as carried out during life cycle assessment (LCA) [26]. LCA is a tool to estimate the potential environmental impact of a product "from cradle to grave" including material extraction, production, transport, construction, operating and management, deconstruction and disposal, recycling and reuse [27].

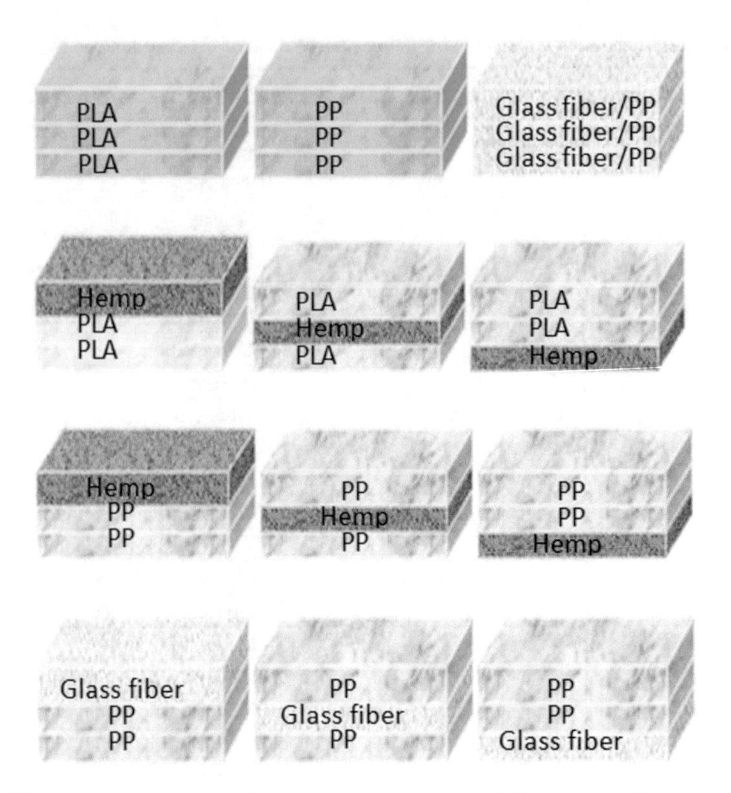

Fig. 1 Schematic diagram of web arrangement in terms of fibre composition and layering sequence for a multilayer sound absorber [4]

LCA determines the environmental impact of a product based on various ecological indicators which include the following:

- cradle-to-grave nonrenewable energy use [27]
- type of waste treatment—based on which the emissions are predicted
- greenhouse gas emission
- ozone precursors
- human toxicity
- acidification
- eutrophication [26]
- ozone layer depletion
- mineral resource depletion
- fossil resource depletion
- waste disposal
- transport pollution [27].

It is not possible to consider all environmental impact types of a product, such as an acoustic textile component, during its full life cycle. Thus, LCAs mainly focus on several environmental indicators while omitting others [26]. Some databases such as Ecoinvent and Eco-indicator provide "eco-profiles" for designers [27]. International Industrial Organization for Standardization (ISO) also developed a standard to describe the principles and framework of LCA ISO 14040:2006 Environmental management—Life cycle assessment—Principles and framework.

Ecoinvent, an independent not-for-profit association comprising of five Swiss institutes, provides LCA and life cycle management data and service to industry, consultancies, public authorities, and alike. It covers more than 10,000 Life cycle inventory databases in various areas such as energy supply, transport, construction materials, and others [28]. The environmental indicators that are taken into account by Ecoinvent include global warming potential, acidification, energy demand and nonrenewable energy consumption among others [27].

Eco-indicator 99 classifies harms done to environment in three groups: human health (climate change, ozone layer depletion, nuclear radiation, as well as carcinogenic effects and respiratory disorders), eco system quality (ecotoxicity, acidification, eutrophication and land use), and resources (mineral resource depletion, fossil fuel depletion). Eco-indicator 99 use points for LCA assessment where 1 point corresponds "*to one thousandth of the yearly environmental load of one average European inhabitant.*" Materials; manufacturing, transport, energy generation processes and disposal scenarios are evaluated using this eco points by Eco-indicator 99 [29].

3 Environmental Concerns of Manufacturing

Whereas the standards have been enhanced substantially in the manufacturing industry due to advancement in science and technology, it becomes very important to balance the development with responsible environmental practices [4]. Within this scope, sustainable development comes into effect which is defined as

> Development that meets the needs of the present without compromising the ability of future generations to meet their own needs.

in the Report of the World Commission on Environment and Development (the Brundtland Commission): Our Common Future [30] which is also known as the Brundtland Report [27].

Among different manufacturing industry segments, the textile industry is among the longest and most complex industrial chains. It has a fragmented and heterogeneous structure composed mainly by SMEs, and consists of various subsectors, starting the production cycle from raw material production (synthetic and regenerated fibres) to semi-processed (yarns, fabrics and finishing processes) and end products (clothing, home textiles and industrial textiles) [31].

Similar to other industrial areas, the textile industry includes activities that may have severe impact on the environment. These include but not limited to wet treatments (dyeing, finishing), and laminating and other joining processes [32].

3.1 Impact of Textile Processes on the Environment

The great amount of effluent water, with its chemical load is the major environmental issue related to the textile industry. Air pollution, solid wastes, and unpleasant odours are other sources of environmental concerns together with energy [33] and water consumption [34].

The impact of textile industry is started to be felt at the first step of the supply-chain: fibre production. Even if natural fibres are used for manufacturing acoustic textiles, cultivation of fibre plants, especially cotton, incorporates pesticide use, and application of artificial irrigation methods [35]. Then, water extraction of bast fibres lead to high chemical oxygen demand (COD) loads in the effluent. On the other hand in the course of synthetic and regenerated fibre production emissions of aromatic diisocyanates, nitrogen dioxide (NO_2), sulphur and other volatile organic compounds and heavy metals may occur based on the fibre type [34].

The unwanted substances found on the raw materials–fibres–such as impurities, biocides (for natural fibre), preparation agents, spinning lubricants and sizing agents are removed in the course of pretreatment processes and end up in effluent water in the form of organic substances that are hard-to-biodegrade like mineral oils, together with polyaromatic hydrocarbons, alkyl phenol ethoxylates (APEO), and biocides, which are hazardous compounds, and chemical oxygen demand (COD) loads. If heat setting is applied, in this case these substances become airborne and cause air pollution [31].

Desizing of cotton and cotton blend products result in very high COD loads in the effluent water [31]. Scouring results in COD load, high alkalinity and increase in temperature of the surface water where the effluent is discharged [34].

Sodium hypochlorite and chlorine bleaching give rise to adsorbable organic halogens (AOX) in the waste water wherein that of the latter is much lower than the first one. On the other hand, the problem associated with hydrogen peroxide bleaching is due to the use of strong compounding agents like stabilizers. As for mercerization process, if the rinsing water is not reused the effluent water will carry very high alkalinity levels [36].

Most textile products, including acoustic textiles are subjected to certain colour treatments such as dyeing. The dyeing processes result in emissions, most of the times, to water. Dyeing effluents contain pollutants originated from the dyestuff (metal, colour, etc.), auxiliaries in the dye formulation and those used in dyeing processes together with contaminants present on the fibre. While dyeing process, itself, leads to very high COD levels, rinsing after dyeing results in high water consumption. In terms of printing, residues of the printing paste, waste water from washing and cleaning operations and volatile organic compounds from drying and fixing processes constitute the general emission sources [31].

For finishing processes, as subsequent rinsing is generally not necessary, once curing has been carried out, air emissions become more important than water emissions and water consumption. During the drying and curing operations the emissions to air take place to the extent determined by the volatility of finishing agents as well as that of the compounds carried over from previous processes [31].

Resulting in such a heavy burden on the environment, it has become obligatory to take precautions such as supplying improved knowledge of the process inputs and outputs to the management and staff, defining well-documented procedures at each level of the production, improving the chemicals used in quality and quantity, optimizing water and energy use as well as proper handling of the waste waters [31] in the course of manufacturing acoustic textiles as well as other textile products.

Within this concept, the requirements for a textile material to be considered as ecological textiles/ecotextiles/sustainable textiles has been listed by Rose [37] as follows:

- All materials and process inputs and outputs are safe for human and ecological health in all phases of the product life cycle.
- All energy, material and process inputs come from renewable or recycled sources.
- All materials are capable of returning safely to either natural systems or industrial systems.
- All stages in the product life cycle actively support the reuse or recycling of these materials at the highest possible level of quality.
- All product life cycle stages enhance social well-being.

(http://coralrose.typepad.com/my_weblog/2008/05/sustainable-tex.html)

Besides, the manufacturing industry is prompted to select materials and processing techniques with less impact on the environment by a number of regulations. Some of them that are related with manufacturing of acoustic textiles are listed below.

3.2 Some Legislation Concerning Acoustic Textiles

A relevant environmental legislation is the Directive 2010/75/EU on Industrial emissions (Integrated pollution prevention and control (IED)) which goals minimizing industrial emissions that are harmful in terms of human and environmental ecology via ensuring the application of the best available techniques in wide industrial segments [38]. The best available technique entails the criteria of using low-waste technology with less-hazardous substances and by incorporating improved waste recovery and recycling that have met previous success in industry level [36].

With regard to textile wet treatment processes, for pretreatment (washing, bleaching, mercerization, etc.) and dyeing facilities possessing capacities exceeding 10 ton per day to operate, an environmental permit is necessary. The conditions of this permit should comply to Best Available Techniques (BAT) as previously stated [38].

The new REACH Regulation (Regulation on Registration, Evaluation, Authorisation and Restriction of Chemicals) that came into force in 2007 also affects the textiles industry which is a huge consumer of chemicals. REACH aims at ensuring that the human health and the environment are protected from the potential harms of chemicals by evaluating and managing the risks, supplying safety information for the users, and taking additional executional precautions for very hazardous substances [33, 38].

Among regulations related to automobiles wherein noise control is a major issue, European end-of-life vehicle (ELV) legislations necessitate the fraction of an end-of-life vehicle that is deposited in landfills to be decreased to 5 % by the year 2015 with a minimum reuse and recycling rate of 85 % and reuse and recovery rate of 95 %. With the guidance of the targets set by this legislation, the average reuse and recycling, reuse and recovery rate has been increased from 78 and 81 % in 2006 to 84 and 88 % in 2011 [39]. According to 2011 figures, among 27 European Union member states, 4 reached 2015 targets for reuse recovery and 11 for reuse and recycling. The ELV legislation resulted not only in significant reduction of the environmental impacts of ELVs but also prompted innovation as some car manufacturers developed new designs for their vehicles which allow high recycling and recovery rates after use by integration of higher amount recycled materials [40].

To give another example related to civil engineering field, many municipalities in Italy have introduced Building Regulations which encourage use of materials with lower impact on the environment in new building constructions by suggesting decrease in construction taxes. A list of materials which should be abstained from, including mineral fibres, is also included in these regulations [27].

3.3 Material Selection in Noise Control Elements

Protection of the environment should not be realized with compromising industrial progress, when adopting environmentally sustainable components as raw materials [4]. The balance of environment protection and technology development can be achieved using constituents that lead to minimal environmental impact while possessing performance characteristics comparable to conventional synthetic materials at the same time [41]. Whereas textiles in acoustical applications may be found as natural and recycled fibres, and nonwovens; glass fibre mats and foams are commonly used conventional components in acoustic materials (Table 3).

3.3.1 Glass Fibre and Mineral Wool

Glass fibre has been the major sound absorber material for the past half century and found use in ceiling tiles, duct liners; interior lining in buildings, vehicles and aircraft [21, 44, 45]. Used as the most common acoustic absorber material, glass fibre mats have been the subject of extensive research studies related to sound absorption.

However, glass fibre and mineral wool can be problematic in terms of human health and environmental protection. Glass fibres and mineral wool fibres can go into the lung alveoli when inhaled and can cause skin irritation when touched. Thus, these materials are a concern in terms of health safety and should be properly sealed to protect the final users [27, 4] and precautions should be taken when handled, to protect workers. Mineral wool is among the construction materials that have the greatest environmental impact [29]. Moreover, glass fibre leads to high greenhouse gas emission [27] and it cannot be recycled [46]. The high density of glass fibre (2.5 g/cm^3), compared to natural fibres which is around 1.5 g/cm^3 results, in higher energy consumption and emissions if used for noise control in transportation [4, 42].

3.3.2 Foams

Another type of acoustic materials that textiles compete with is foams. As mentioned before, sound absorbers can be generally classified as cellular, granular and fibrous materials, and foams fall into the first category: cellular materials. Among different types of foams, polyurethane foam enjoys the most common use as acoustic absorption material in spite of its high flammability and toxic combustion byproduct emission during burning [32, 4].

Polyurethane foam enjoys endurance against dynamic loads, good compressibility and high recovery rates [47]. Polyurethane foam is produced by two main components, an isocyanate and a polyol, as well as additives and catalysts. Vapour and aerosols emitted during mixing of the chemicals and curing of polyurethane foam that include aromatic diisocyanates can lead to respiratory problems including asthma, lung damage; irritation of respiratory system, skin and eye; and even cancer [27]. The Environmental Protection Agency (EPA) issued personal safety guidelines for people who work in environments exposed to vapour of polyurethane [48]. Moreover, polyurethane foam is flammable, and when burned it may emit toxic fumes. As polyurethanes are thermoset materials that means once cured they cannot be remelted into new shapes, they cannot be recycled [49]. However, shredded polyurethane foam waste can be reused in different applications [50]. The energy consumption of polyurethane foam is also much higher than natural fibres [27].

Another issue related to use of polyurethane foam is, the flame lamination method, which has widespread use in joining the foam layer to the adjacent layers in seat, headliner, and door casing in vehicle production worldwide. It includes subjecting the foam surfaces to gas flame and let the molten foam to bind itself to the next layer [21]. However, this flame lamination method has some critical impact on the environment. When thermally degraded, foam, including blowing and flame retarding agents as well as other additives, may emit hazardous gases such as cyanide, hydrogen chloride and hydrogen bromide [15].

As a result of the mentioned concerns, among petro-based polymers polyurethane foam has been rated to have one of the greatest environmental impact in terms of production as well as after-use stages by Eco-indicator 99 [29].

Another type of foams which found use in acoustical materials is melamine foams. Similarly to polyurethane foam their production is based on petro-chemicals [27] and they are thermoset; hence, not recyclable, whereas the flammability is not the case such as polyurethane foams [51]. Whereas the environmental impact of melamine foams is lower than polyurethane foams, they are not as cost efficient and tear resistant as polyurethane foams [52].

3.3.3 Nonwovens

Nonwovens stand as promising environmentally friendly alternatives to the conventional sound absorbers including glass fibre mats and polyurethane foams thanks to their production methods that inflict less damage to the environment. Here nonwovens refer to fibrous structures made of textile fibres including natural (fibres of plant and animal origin), synthetic (polyester, polypropylene, polyamide, acrylic, elastane) and man-made cellulosic (viscose, modal and lyocell) excluding glass fibre [53, 4].

Nonwoven fabrics are defined by Association of the Nonwoven Fabrics Industry (INDA) as "*sheet or web structures bonded together by entangling fibre or filaments (and by perforating films) mechanically, thermally, or chemically*" (2015). The global production figure was reported to be 33.1 billion USD dollars in 2013, and an increase to 47.8 billion dollars in 2018 is expected with a projected annual growth rate of 7.5 % [54]. Nonwovens have found use as noise control applications in acoustic ceilings, wall claddings and acoustic barriers [4], noise absorbers in passenger vehicles among others. Nonwovens enjoy relatively low manufacturing costs. Nonwovens may also inflict less harm to the environment compared to conventional polyurethane foams and glass fibres in terms of production and after-use stages of their life cycle [55, 4].

Although nonwovens, produced from conventional textile fibres, are more environmentally sustainable compared to polyurethane foams or glass fibres in various ways; that is not enough to maintain commercial sustainability of nonwovens. As most sound absorbers carry out multiple tasks other than noise control, such as thermal insulation (Acoustical surfaces) and comfort for the case of polyurethane foams. Thus, nonwovens aimed for foam replacement should also endure dynamic loads, possess good compressibility and high recovery rates [47].

3.3.4 Natural Fibres (Plant and Animal Fibres)

Here, natural fibres refer to plant and animal fibres as stated in European Commission decision [53]. Other than conventional man-made materials, natural fibres also have found use in noise control elements. Whereas noise control capability of some natural fibres is under search, some have already been utilized commercially in acoustical materials. Coir, hemp and kenaf fibres are among the commercially used natural fibres in acoustical materials. It is preferred that the fibres are procured from local suppliers in order to cut down on the effect of transport on the environment [27].

Natural fibres depend on renewable resources and biodegrade upon disposal. During the growth period, plant fibres have affirmative ecological impact in terms of CO_2 emission—as they sequester rather than emit CO_2 [27]. They also have low embodied energy compared to petro-based materials and fibres [43]. Despite the fact that natural fibres are based on renewable resources and possess biodegradability, they are not automatically environmentally friendly. Immense amounts of water, fertilizers and pesticides are used during cultivation of natural fibres [56]. The more they are subjected to treatments, the greater their impact on the environment in terms of energy and resource consumption and emissions [27]. Natural fibres should not be considered environmentally sustainable even when no pesticide or other chemicals are used. Extraction of bast fibres by water retting results in high chemical oxygen demand (COD) and total organic carbon (TOC) in effluent water. Bast fibres can only be certified with the Nordic ecolabel SWAN, if the effluent from water retting process is treated and COD and TOC is reduced by 75 % for hemp and 95 % for flax and other bast fibres [34].

The lower density impart natural fibres environmental and economical advantages by cutting down on emissions and energy consumption especially if used in transportation [4]. They do not lead to fogging and do not endanger workers' health and safety, as glass fibre, polyurethane foam and other petro-based materials do. Cost effectiveness and good thermal insulation properties are among the other superior features of natural fibres [43].

In terms of performance characteristics, these fibres are more vulnerable to fungal and microbial attack and possess lower flame retardancy [27] and inferior mechanical strength compared to glass fibres. They are also hydrophilic which may have an adverse effect on its performance characteristics during use. Moisture intake leads to swelling of the fibre and result in deterioration of the dimensional stability of the noise control element. Thus, the natural fibres should undergo modification processes to impart hydrophobicity [42] or should be isolated from moisture sources. When using biomaterials, mold growth becomes a concern. So, the materials must be pretreated with human ecology-safe fungicides. Unflammability is also a major issue. Safe fire retardants such as a boron-based fire retardant application of fibres have found commercial application [57].

As most of the plant fibres start to decompose around 200 °C, the production and service environments should not entail temperatures higher than the mentioned temperature. To avoid photo degradation, they should not be exposed to sun light for prolonged durations [42].

While presenting irregular cross-sections compared to well-defined shapes of petro-based fibres, the plant fibres possess higher diameters compared to high finenesses that can be reached with conventional synthetic fibres [43], especially glass fibre as seen in Fig. 2. The coarseness is a drawback for acoustical absorption, in contrary to irregular cross-section [4]. The difficulties in establishing a theoretical model based on microstructural parameters have been mentioned in the design section. Using natural fibres results in an additional difficulty, to calculate the mean fibre cross-sectional diameter or surface area of a fibrous assembly, due to high irregularity [43] high variation [4].

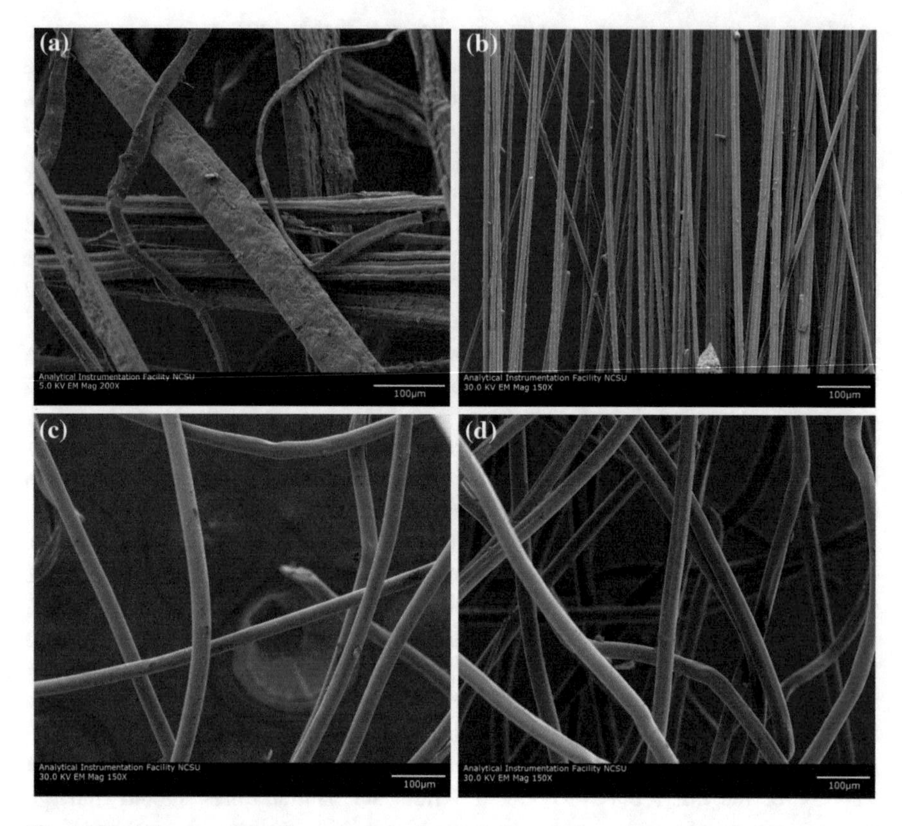

Fig. 2 Scanning electron microscopy of **a** hemp, **b** glass fibre, **c** PLA (poly (lactic acid)), and **d** PP (polypropylene) fibres at 30.0 kV. The magnification of hemp fibre is × 200 and the others × 150 (From Yilmaz et al. [13]. With Permission from Wiley)

High variability is a major concern related to natural fibres. There are a number of uncontrollable factors which impart variation on natural fibres, namely fibre crop species, farm location, climate, time and method of harvesting, and alike [42].

Yilmaz et al. [55, 58] compared sound absorption (Fig. 3a) and sound barrier properties (Fig. 3b) of fibrous materials made from hemp, poly (lactic acid) and polypropylene which were manufactured using the same production parameters. Comparable noise control capability of natural fibres compared to the man-made fibres was reported.

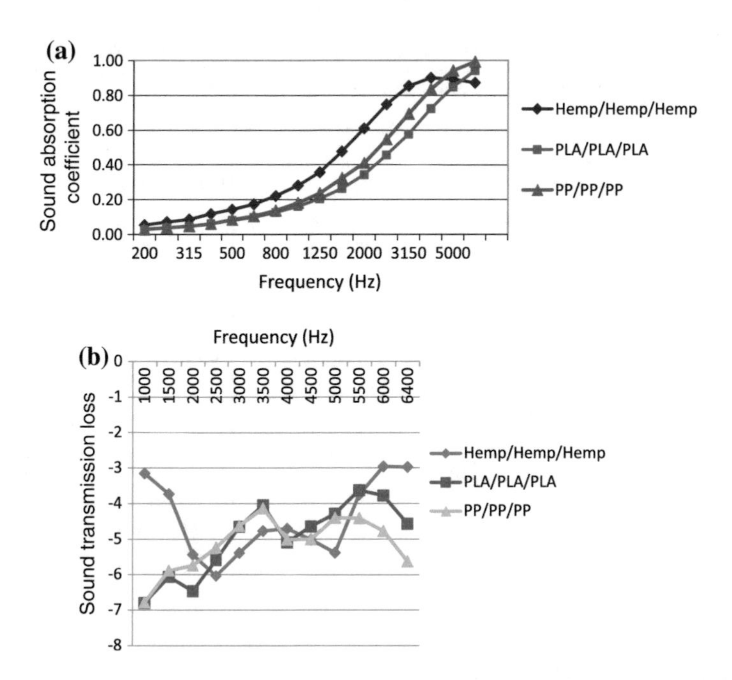

Fig. 3 Comparison of three-layer nonwovens of hemp, poly (lactic acid) (PLA) and polypropylene (PP) in terms of **a** sound absorption [58] and **b** sound transmission loss [55]

Summary of Research Studies Related to Use of Natural Fibres in Acoustical Materials

There has been a variety of studies relating to use of biodegradable fibres and/or polymers for noise control applications which incorporate conventional and exotic plant fibres, animal fibres, and engineered compostable polymer fibres [25].

Within this concept, Nick et al. [14] studied sound absorption capabilities of polypropylene blends with cotton, flax and hemp. They reported superior sound absorption of the cotton—polypropylene blend.

Yilmaz et al. [12] compared the noise control capabilities of sandwiched composite structures of hemp and poly (lactic acid) (engineered compostable polymer based on corn) with that of polypropylene and glass fibre. They reported close results from the sound absorption of the biodegradable and the conventional fibrous assemblies (Fig. 4). Yilmaz et al. also studied the effects of fibre type, and layering sequence [13], compression [22], alkalization [23] and thermal treatment [24] on sound absorption and sound transmission loss [55, 58]. They built mathematical models to predict sound absorption coefficient based on material and treatment parameters.

Jayaraman [44] developed nonwoven sound absorbers including various grades of polyester fibres and studied the influence of kenaf fibre addition on the sound absorption of these nonwovens. They reported adverse effect of kenaf fibre, which

Table 3 Comparison of merits and demerits of some materials used in noise control elements

Natural fibres	Glass fibres	Polyurethane foam	Nonwovens[a]
Pros			
Cost efficient	Fine fibres	Cost efficient	Cost efficient
CO_2 neutrality	High thermal durability	Light weight	Low density
Renewable resource	Hydrophobic	Hydrophobic	No wear on equipment
Low density	Resistance against fungi and microbial attack	Resistance against fungi and microbial attack	No risk for health or safety
Thermal insulation	Not flammable	Resilience	Lower emission of hazardous gases during production and processing stages compared to PU foam
No wear on equipment			
No risk for health or safety			
No fogging			
Irregular cross-section			
Cons			
Flammable	Irritation for skin	Risk for health and safety	Low resilience
Prone to fungal and microbial attack	Risk for health and safety	High flammability	Depletion of fuel resources based on fibre content
Coarse	Abrasion in production equipment	Emission of hazardous gases during production and processing stages	
Low thermal durability	High density	Depletion of fuel resources	
High variation	Depletion of fuel resources	CO_2 emission	
Moisture absorption	CO_2 emission		

[a]Nonwovens refer to fibrous structures made of textile fibres, natural, synthetic or man-made cellulose, excluding glass fibre

Note Table has been arranged by the author based on Yilmaz [42], Yilmaz et al. [12], Berardi and Iannace [43] and Asdrubali et al. [27]

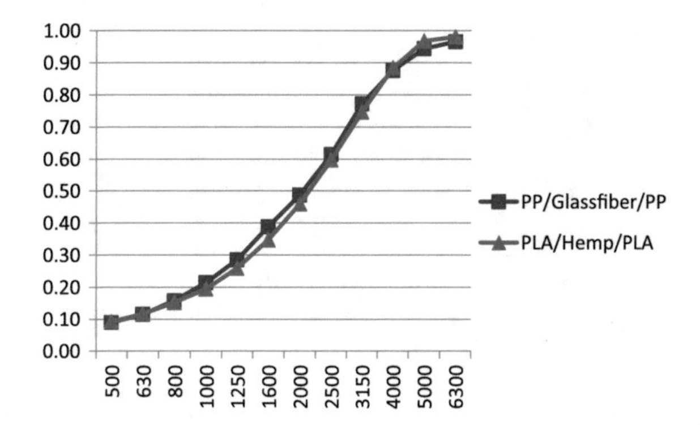

Fig. 4 Comparison of sound absorption of three-layer nonwovens of PP/Glass fibre/PP layers versus PLA/Hemp/PLA [12]

may be due to its high linear density, on sound absorption, which is less pronounced at high frequencies.

Parikh et al. [8] produced different sound absorber materials from kenaf, jute, flax, reclaimed cotton with recycled polyester and off-quality polypropylene fibres to be used in automotive floor coverings.

Huda et al. [59] developed light-weight unconsolidated polypropylene web composites with bamboo strips and jute fibres by compression molding. They found the noise reduction performance of the composite containing bamboo strips to be superior to the other one.

Markiewicz et al. [60] also produced polypropylene composites incorporating different lignocellulosic fillers of hemp, rapeseed straw, beech and flax fibres and determined their sound absorption coefficient within the frequency range of 1000–6500 Hz. While hemp addition was reported to increase in noise reduction over for frequencies over 3000 Hz, the other fillers affected sound absorption positively between 3000 and 4000 Hz frequencies.

Brencis et al. [61] produced gypsum foam reinforced with hemp fibres and determined the sound absorption characteristics. They produced gypsum foam from gypsies rock (gypsum), which is a natural resource and already found commercial use in building acoustical materials [16]. Gypsum possesses high fire resistance—ranking in the fire reaction class A according to with the European regulations. Fragility is its drawback and necessitates use of reinforcement. Brencis et al. [61] reported positive effect of hemp fibre addition on noise reduction capability which is more pronounced for higher fibre lengths (5–10 mm) compared to lower fibre length (2.5–5 mm).

There are relatively fewer studies pertaining to noise control elements including fibres of animal origin. Among them, Ballagh [62] studied the noise control performance of woollen materials to replace mineral—fibre mats. Berardi and Iannace [43] also studied sound absorption of sheep wool (Fig. 5). In another study, Huda and Yang [63] studied the sound absorption capacities of PP composites with ground chicken quill and jute fibre.

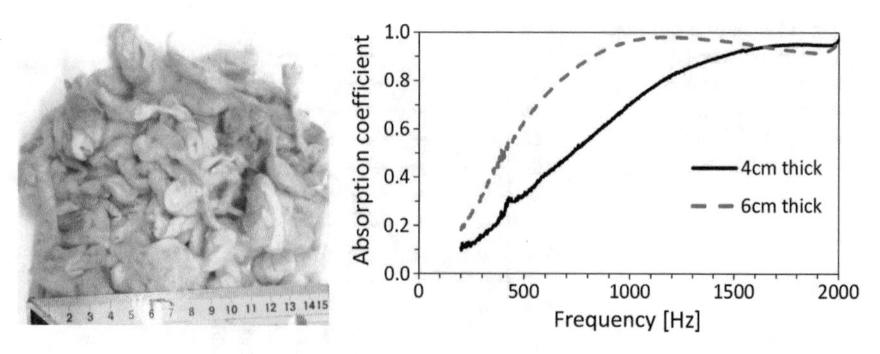

Fig. 5 Absorption coefficient of sheep wool 4 and 6 cm thick (Berardi and Iannace [43]. With Permission from Elsevier)

Biodegradable Versus Recyclable

Yilmaz [4] conducted an expert evaluation study as a new product forecasting method related to some novel sound absorbers targeting use in passenger vehicles. The sound absorbers were produced from fibres including hemp, glass fibre, polypropylene and polylactic acid. In the conducted expert evaluation study, it was stated that biodegradability is overrated in the general community. Recycling of the material in the after-use stage maybe more preferable for environmental and economical point of view.

3.3.5 Recycled Materials

There is an interest toward developing acoustic materials from various recycled materials including wastes of foam, tyre [19], rubber [64, 4], plastic, textile and metal. Acoustical panels made of post-consumer and post-industry recycled newspaper, post-consumer recycled denim have already found commercial application [57]. Some ecolabels determine the minimum recycled content in a product. As an example, Nordic ecolabel requires 70 % of wood fibre content in construction panels should be recycled or byproducts of wood processing industry such as shavings or sawdust [65].

When developing acoustic material incorporating recycled constituents, the noise reduction behaviour of the material can be fine-tuned by blending particles of different sizes. A proper binder might be necessary to provide integrity of the material [27]; on the other hand, when using recycled materials of thermoplastic origin, binder use might not be necessary.

In a relevant study, recycled polyester nonwovens were evaluated as sound absorbers by Lee and Joo [66] and the effects of certain parameters including fibre fineness and cover screen inclusion have been searched. In another study, Lou et al. [67] produced sound absorbers from recycled polyester and polypropoylene

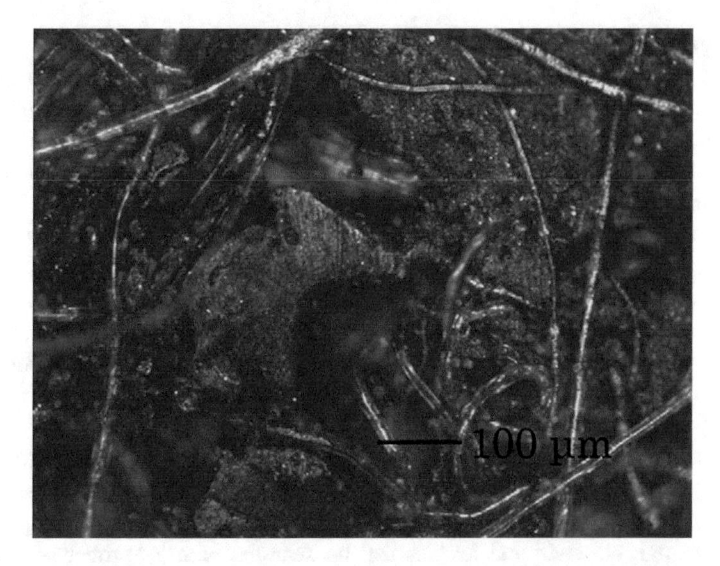

Fig. 6 Microphoto of the sound absorber produced from cord fibres and ground rubber particles obtained from end-of-life tyres (From Maderuelo-Sanz et al. [68]. With Permission from Elsevier)

nonwoven selvages by compression molding and investigated the effects of production parameters as well as sawdust addition on the sound absorption performance.

In a different work, Maderuelo-Sanz et al. [68] produced sound absorbers from cord fibres and ground rubber particles obtained from shreddings of end-of-life tyres by compression molding. They evaluated the effects of acrylic resin inclusion, thickness, density, pressure and temperature on sound absorption. The microstructure of the absorber is shown in Fig. 6. There are also other studies related to use of shredded foam waste [27] or pellets from waste plastic bottles [69] which fall outside the 'acoustic textiles' concept.

3.4 Ecolabels

Emergence of ecolabels is rooted in the growing concerns of the community for the past few decades regarding to environmental protection and endeavour of businesses to transfer these concerns into market advantage. Within this concept, the businesses started to market their products and services with 'green claims' by qualifying them as ecofriendly, natural, recyclable, etc. [70]. However, chances are these claims are not accurate (greenwashing). Even if the green claims are arranged based on good will, they can never guarantee reliability. Thus, to avoid misleading of consumers, ecolabels came into existence. Ecolabels are "*independent and reliable labels that consider the life-cycle impact of products and services*" [71].

The ecolabels are issued by independent third parties, not under the influence of the business who strives to get certification. They are based on sound scientific evidence [71]. Ecolabels cover specific product categories some are which are related to acoustical materials.

Even though ecolabels are rarely available for innovative or specialist goods [71], some noise control element producers provide eco labels related to the effect of their material on human health and the environment [16]. Acoustic material producers who manufacture building noise control elements including wood material may show FSC (Forest Stewardship Council) certificate [16, 72]. Acoustic products intended to use in buildings may be certified by BRE Global, which measure environmental impacts of construction materials and products throughout the life cycle including extraction, processing, use, and maintenance and after service life stages [73].

Acoustic textiles, which might be classified under the interior textiles class including fillings, linings, paddings, membranes, and coatings made of fibres, can be certified by European Ecolabel through the product group "textile products" [53]. Although conventional textiles can be awarded with a number of ecolabels including European ecolabel (flower), Nordic ecolabel (swan), Bra miljöval (falcon) and Eoko-Tex [74, 70], the options are limited for acoustic textiles. Notwithstanding, the number of product groups covered by the ecolabels are increasing and open for suggestions as stated in the official web site of Nordic Ecolabeling [65].

4 Recycling Issues

"Reduce, reuse, recycle." These environmentalist watchwords draw a guideline for acoustic textile producer who want to contribute to preserving of resources and protection of any type of pollution. These 3R's tell to reduce consumption of resources and waste generation, reuse the product at the end of service life in the same form if possible, if not possible, then recycle it and use it in a different form [32]. It sounds nice, but all materials do not offer the same easiness for recycling.

The content of the acoustic material may be either homogeneous or blend. In the case of mixed materials, separating and sorting of individual components for subsequent recycling processes is a problem. This might become easier if the label of each component is provided. If material information is not available, time-consuming characterization tests become necessary including chemical analyzes, microscopic investigation, burning tests, melting point determination, etc. Thus, proper labelling of each component is of paramount importance [15].

On the other hand, if the material is produced from a single polymer, the sorting process will be needed. In the case of mixed materials, neither natural fibre reinforced composites nor glass fibre reinforced composite structures promise ease for recycling. The commonly used method to fabricate single-polymer composites includes using grades of the same polymer with different crystallization/alignment degree, and hence, melting points. The one with the lower melting temperature acts as the thermal adhesive/binder/matrix whereas the other one provides strength [75]. Polyester or propylene may be options for single-polymer fibrous materials [15].

In an automotive-related example, the passive noise control elements take place in automobile carpets present a complicated structure including a sintered layer, high density layers, a secondary backing and an acoustic damper. This complicated structure is constituted by a variety of materials including polyester, polypropylene, ethylene vinyl acetate (EVA), ethylene propylene diene terpolymer (EPDM), bitumen and mineral filler. Due to the difficulty in separating and sorting of the contents, these materials may end up in incinerators or landfill sites rather than being recycled [15].

Research efforts have taken place to produce single-polymer carpet and noise control systems. An example of this is a 100 % polyester carpet comprising of polyester piles, spunbond nonwoven polyester primary backing, polyester hot melt adhesive and a needle-punched nonwoven polyester secondary backing developed by Williams et al. [76] and filed under US patent application number US 20140272262 A1.

Another option for making recycling possible and easier to use is adopting easily separable mixed materials. In the case, where use of single-polymer composites is not a viable option, mixed materials which allow easy detachment may be preferred. The list below is arranged in an ascending order based on ease of recycling [77].

- Mixed material—permanently fixed
- Mixed material—separable joining
- Single-material composite
- Single material

Whether the components are retrieved from homogeneous or mixed materials, they should be investigated in terms of their chemical finishes which may induce odor, fogging or stiffness in new goods [15]. The materials should also not contain hazardous materials; this may pose significant difficulties in the recycling stage [78].

5 Concerns at the End-of-Life

In terms of environmental friendliness, the acoustic material should be able to preferably biodegrade or recycled at the end of useful service life. If not possible it should be incinerated providing good calorific value. None of these end-of-life scenarios should take place by emitting hazardous substances to air, soil or water [41].

In case of incineration, it is preferred that the material can be incinerated with a good calorific value. Whereas polypropylene and polypropylene can be incinerated with reasonable energy yield, nylon yields lower energy [29]. The toxic gas release from polyurethane foam outweighs the positive effect of energy production during its incineration [29, 32].

For landfill disposal, components that do not react with other materials pose lower threat to the environment, whereas the ones that react and cause emissions of greenhouse gases like CO_2 and methane have higher impact. Besides, voluminous substances, which occupy larger volume per unit mass is also disadvantageous as they play a greater role in landfill space depletion [29].

6 Conclusion

Textiles have begun to find use in acoustical applications for the past few decades. In order to consolidate the position of textiles as noise control materials, their performance characteristics should be further enhanced to a level comparable to conventional acoustic materials, the most common ones of which are glass fibre mats and polyurethane foams.

The same recent history has also witnessed substantial growth in public concern related to environmental effects of industrial progress. As much as this situation imposes a burden on textile producers, it also opens a new battlefield against acoustic materials, where textiles can prevail. Conventional acoustic materials, mineral wool, or polyurethane foam among others are under line of fire due to their adverse effects on the ecosystem as well as on human health. This situation can offer business advantage to textile producers, provided that the damage inflicted on the environment throughout the whole life cycle of the textile product is minimized and the functional properties are improved.

In this sense, this chapter has considered acoustic textiles from an ecological point of view. Their design stage based on the functional and environmental requirements has been reviewed. The effects of manufacturing stage of acoustic textiles on the environment have been investigated. Different materials used for sound absorption have been screened and compared. Recycling issues and end-of-life concerns have been referred to.

References

1. Na Y, Lancaster J, Casali J et al (2007) Sound absorption coefficients of micro-fiber fabrics by reverberation room method. Text Res J 77(5):330–335. doi:10.1177/0040517507078743
2. Findeis H, Peters E (2004) Disturbing effects of low frequency sound immissions and vibrations in residential buildings. Noise Health 6(23):29–35
3. Haapakangas A, Helenius R, Keskinen E et al (2008) Perceived acoustic environment, work performance and well-being-survey results from Finnish offices. In: Anonymous 9th international congress on noise as a public health problem, Foxwoods, CT
4. Yilmaz ND (2009) Acoustic properties of biodegradable nonwovens. North Carolina State University
5. EPA (2012) Noise pollution. Available via. http://www.epa.gov/airprogm/oar/noise.html. Accessed 12 June 2015

6. Al-Nawafleh MA (2005) Social-economic and ecological criteria of noise control. Am J Appl Sci 2(2):496–498. doi:10.3844/ajassp.2005.496.498
7. Kaarlela-Tuomaala A, Helenius R, Keskinen E et al (2012) Effects of acoustic environment on work in private office rooms and open-plan offices—longitudinal study during relocation. Ergonomics 55(12):ebi
8. Parikh DV, Chen Y, Sun L (2006) Reducing automotive interior noise with natural fiber nonwoven floor covering systems. Text Res J 76(11):813–820. doi:10.1177/00405 17506063393
9. Wilson A (2006) Engineered nonwovens used for automotive acoustic insulation. Tech Text Int 15(8):14
10. Hodge A (2012) Noise pollution acts as a biodiversity filter. Available via. http://www.ecology.com/2012/01/26/noise-pollution-biodiversity/. Accessed 15 June 2015
11. Francis CD, Kleist NJ, Ortega CP et al (2012) Noise pollution alters ecological services: enhanced pollination and disrupted seed dispersal. Proc Biol Sci/R Soc 279(1739):2727–2735. doi:10.1098/rspb.2012.0230
12. Yilmaz ND, Banks-Lee P, Powell NB (2009) Hemp-PLA needle-punched nonwoven composite as a 'green' sound absorber. In: Anonymous needle punch 2009. INDA, SC
13. Yilmaz ND, Banks-Lee P, Powell NB et al (2011) Effects of porosity, fiber size, and layering sequence on sound absorption performance of needle-punched nonwovens. J Appl Polym Sci 121(5):3056–3069
14. Nick A, Becker U, Thoma W (2002) Improved acoustic behavior of interior parts of renewable resources in the automotive industry. J Polym Env 10(3):115–118. 1021124214818
15. Mukhopadhyay S, Partridge J (1999) Automotive textiles. Text Progress 29(1):1–125. doi:10.1080/00405169908688876
16. Anonymous (2013) Product overview. Available via. http://www.topakustik.ch/en/product-overview/. Accessed 15 June 2015
17. Attenborough K, Ver IL (2006) Sound-absorbing materials and sound absorbers. In: Ver IL, Beranek LL (eds) Noise and vibration control engineering, 2nd edn. Wiley, NJ, p 277
18. Lehner S (2005) European fire classification of construction products, new test method "sbi", and introduction of the european classification system into german building regulations. Otto-Graf J 16:166
19. Cox TJ, D'antonio P (2004) Acoustic absorbers and diffusers. Taylor & Francis, New York
20. Rossetti S, Gardonio P, Brennan MJ (2005) A wave model for rigid-frame porous materials using lumped parameter concepts. J Sound Vib 286(1):81–96. doi:10.1016/j.jsv.2004.09.034
21. Fahy F (2001) Foundations of engineering acoustics. Academic Press, San Diego
22. Yilmaz ND, Michielsen S, Banks-Lee P et al (2012) Effects of material and treatment parameters on noise-control performance of compressed three-layered multifiber needle-punched nonwovens. J Appl Polym Sci 123(4):2095–2106
23. Yilmaz ND, Powell NB, Banks-Lee P et al (2012) Hemp-fiber based nonwoven composites: effects of alkalization on sound absorption performance. Fiber Polym 13(7):922. doi:10.1007/s12221-012-0915-0
24. Yilmaz ND, Powell NB, Banks-Lee P et al (2013) Multi-fiber needle-punched nonwoven composites: effects of heat treatment on sound absorption performance. J Ind Text 43(2):246
25. Yilmaz ND, Powell NB (2015) Biocomposite structures as noise control elements. In: Thakur VK, Kessler M (eds) Green biorenewable biocomposites: from knowledge to industrial applications. Apple Academic-CRC Press, New York, p 198
26. Patel M, Narayan R (2005) How sustainable are biopolymers and biobased products? The hope, the doubts, and the reality. In: Mohanty AK, Misra M, Drzal et al (eds) Natural fibers, biopolymers, and biocomposites. CRC Press, p 854
27. Asdrubali F, Schiavoni S, Horoshenkov K (2012) A review of sustainable materials for acoustic applications. J Building Acoust 19(4):312. doi:10.1260/1351-010X.19.4.283
28. Ecoinvent centre (2015) The ecoinvent database. Available via. http://www.ecoinvent.org/database/database.html. Accessed 8 July 2015

29. Eco indicator (2000) Eco indicator 99 manual for designers. Available via. http://www.pre-sustainability.com/download/EI99_Manual.pdf. Accessed 1 July 2015
30. World Commission On Environment and Development (1987) Our common future. Oxford Paperbacks, Brutland
31. European Commission (2013) Reference document on best available techniques for the textiles industry 1-586. European Commission Integrated Pollution Prevention and Control (IPPC). Available via. http://eippcb.jrc.ec.europa.eu/reference/BREF/txt_bref_0703.pdf. Accessed 25 June 2015
32. Fung W, Hardcastle M (2000) Textiles in automotive engineering, 1st edn. Woodhead Publishing, NY
33. European Commission (2013) REACH—registration, evaluation, authorisation and restriction of chemicals. Available via. http://ec.europa.eu/enterprise/sectors/chemicals/reach/index_en.htm. Accessed 20 June 2015
34. Nordic Ecolabelling (2012) Nordic ecolabelling of textiles, hides/skins and leather. Available via. http://www.nordic-ecolabel.org/criteria/product-groups/?p=3. Accessed 8 July 2015
35. Naturskyddsföreningen (2015) Textiles. Available via. http://www.naturskyddsforeningen.se/in-english/good-environmental-choice/textiles. Accessed 8 July 2015
36. European Commission (2010) Directive 2010/75/eu of the European parliament and of the council. Available via. http://eur-lex.europa.eu/legal-content/EN/TXT/HTML/?uri=CELEX:32010L0075&from=EN Accessed 8 July 2015
37. Rose C (2008) Sustainable textiles. Available via. http://coralrose.typepad.com/my_weblog/2008/05/sustainable-tex.html. Accessed 15 June 2015
38. European Commission (2014) Environmental issues. European Commission. Available via. http://ec.europa.eu/enterprise/sectors/textiles/environment/index_en.htm. Accessed 20 June 2015
39. Eurostat (2013) End-of-life vehicle statistics. Eurostat. Available via. http://ec.europa.eu/eurostat/statistics-explained/index.php/End-of-life_vehicle_statistics. Accessed 28 June 2015
40. European Commission (2014) Ex-post evaluation of certain waste stream Directives—Final report. European Commission. Available via. http://ec.europa.eu/environment/waste/pdf/target_review/Final%20Report%20Ex-Post.pdf. Accessed 30 June 2015
41. Mohanty AK, Misra M, Drzal LT et al (2005) Natural fibers, biopolymers, and biocomposites. In: Ak Mohanty, Misra M, Misra M et al (eds) Natural fibers, biopolymers, and biocomposites. CRC Press, FL
42. Yilmaz ND (2015) Agro-residual fibers as potential reinforcement elements for biocomposites. In: Thakur VK (ed) Lignocellulosic polymer composites: processing, characterization and properties. Wiley-Scrivener, New York, p 270
43. Berardi U, Iannace G (2015) Acoustic characterization of natural fibers for sound absorption applications. Building Env xxx:13. 10.1016/j.buildenv.2015.05.029
44. Jayaraman AK (2005) Acoustical absorptive properties of nonwovens. North Carolina State University
45. Tascan M (2005) Acoustical properties of nonwoven fiber network structures. Clemson University
46. Huda MS, Mohanty AK, Drzal LT et al (2004) Physico-mechanical properties of "green" composites from polylactic acid (PLA) and cellulose fibers. In: Anonymous GPEC 2004. Society of Plastics Engineers, MI, p 75
47. Chapman R (2010) Applications of nonwovens in technical textiles, 1st edn. Woodhead Publishing, New York
48. Csanyi C (2015) The environmental impacts of polyurethane foam. Available via. http://education.seattlepi.com/environmental-impacts-polyurethane-foam-6242.html. Accessed 20 June 2015
49. Anonymous (2015) What is PU. Bayer. Available via. http://www.bayermaterialsciencenafta.com/businesses/pur/BayerMaterialScienceNAFTA-OurBusinesses-PolyurethaneSystemsandRawMaterials-WhatisPU.html

50. del Rey R, Alba J, Arenas JP et al (2012) An empirical modelling of porous sound absorbing materials made of recycled foam. Appl Acoust 73(7):609
51. Anonymous (2012) Melamine foam. Clark foam products. Available via. http://www.clarkfoam.net/melamine_foam_products.html. Accessed 8 July 2015
52. Anonymous (2014) Melamine foam vs polyurethane foam. SINOYQX-YarQuenXer. Available via. http://www.sinoyqx.com/blog/2014/01/03/melamine-foam-vs-polyurethane-foam/. Accessed 8 July 2015
53. European Commission (2014) Commission decision of xxx establishing the ecological criteria for the award of the EU ecolabel for textile products. European Commission: 1–5
54. Smithers Apex (2015) Significant growth predicted for global nonwovens market between 2013 and 2018. Available via. http://www.smithersapex.com/news/2013/November/significant-growth-predicted-for-global-nonwovens. Accessed 20 June 2015
55. Yilmaz ND, Banks-Lee P, Powell N (2008) Biodegradable nonwoven composites as noise control elements. In: Anonymous international nonwovens technical conference, September 2008. INDA and TAPPI, Houston, TX
56. Anonymous (2005) Tekstil ve konfeksiyon sektöründe ekoloji ve ekolojik etiketler. İTKİB
57. Anonymous (2014) Soundproofing for green building applications. Available via. http://www.acousticalsurfaces.com/green-building. Accessed June 15 2015
58. Yilmaz ND, Banks-Lee P, Powell NB (2008) The effect of alkalinization on acoustic properties of eco-friendly nonwoven noise control elements. In: Anonymous nonwoven enhancements conference, September 2008. AATCC and INDA, Houston, TX
59. Huda S, Reddy N, Yang Y (2009) Ultra-light-weight composites from bamboo strips and polypropylene web with exceptional flexural properties. Composites Part B Eng 43(3):1658–1664. doi:10.1016/j.compositesb.2012.01.017
60. Markiewicz E, Boryslak S, Paukszta D (2009) Polypropylene-lignocellulosic material composites as promising sound absorbing materials. Polymery 54:430–435
61. Brencis R, Skujans J, Iljins U et al (2011) Research on foam gypsum with hemp fibrous reinforcement. Chem Eng Trans 25:164
62. Ballagh KO (1996) Acoustical properties of wool. Appl Acoust 48(2):101–120. doi:10.1016/0003-682X(95)00042-8
63. Huda S, Yang Y (2008) Chemically extracted cornhusk fibers as reinforcement in light-weight poly(propylene) composites. Macromol Mater Eng 293(3):235–243
64. Hong Z, Bo L, Guangsu H et al (2007) A novel composite sound absorber with recycled rubber particles. J Sound Vib 304(1):400–406. doi:10.1016/j.jsv.2007.02.024
65. Nordic Ecolabelling (2012) Nordic ecolabelling of construction and facade panels. Available via. Accessed 8 July 2015
66. Lee Y, Joo C (2003) Sound absorption properties of recycled polyester fibrous assembly absorbers. AUTEX Res J 3(2):78–84
67. Lou C, Lin J, Su K (2005) Recycling polyester and polypropylene nonwoven selvages to produce functional sound absorption composites. Text Res J 75(5):390–394. doi:10.1177/0040517505054178
68. Maderuelo-Sanz R, Nadal-Gisbert AV, Crespo-Amorós JE et al (2012) A novel sound absorber with recycled fibers coming from end of life tires (ELTs). Appl Acoust 73:408
69. Bratu M, Ropota I, Vasile O et al (2011) Research on the absorbing properties of some new types of composite materials. Revista Romana de Materiale- Romanian J Mater 41:147–154
70. Targosz-Wrona E (2009) Ecolabelling as a confirmation of the application of sustainable materials in textiles. Fibres Text East Eur 4:21–25
71. Anonymous (2009) A guide to environmetnal labels - for Procurement Practitioners of the United Nations System. Available via. https://www.ungm.org/Areas/Public/Downloads/Env_Labels_Guide.pdf/. Accessed 8 July 2015
72. FSC (2015) Certification. Available via. https://us.fsc.org/certification.194.htm. Accessed 20 June 2015
73. Big Room (2015) Big Room 2015 ecolabel index. Available via. http://www.ecolabelindex.com/ecolabels/?st=category,building_products. Accessed 8 July 2015

74. Koszewska M (2011) Social and eco-labelling of textile and clothing goods as means of communication and product differentiation. Fibres Text East Eur 4:20–26
75. Matabola KP, De Vries AR, Moolman FS et al (2009) Single polymer composites: a review. J Mater Sci 44(23):6213–6222. doi:10.1007/s10853-009-3792-1
76. Williams DR, Wallen J, Dawson TG et al (2014) Recyclable single polymer floorcovering article. US20140272262 A1
77. Wang Y (2006) Recycling in textiles, 1st edn. Woodhead Publishing, New York
78. European Commission (2000) Directive 2000/53/EC of the European Parliament and of the Council. Available via. http://eur-lex.europa.eu/legal-content/EN/TXT/?uri=celex:32000L0053. Accessed 10 May 2015

Glossary of Terms

Absorption In acoustics, absorption is the change of sound energy to heat (see sound absorption).

Absorption coefficient The absorption coefficient is a common quantity used for measuring the sound absorption of a material and is known to be the function of the frequency of the incident wave. It is defined as the ratio of energy absorbed by a material to the energy incident upon its surface. It has a value between 0 and 1 and varies with the frequency and angle of incidence of the sound.

Acoustics Acoustics is the interdisciplinary science that deals with the study of all mechanical waves in gases, liquids and solids including topics such as vibration, sound, ultrasound and infrasound.

Acoustic absorption Acoustic absorption or sound absorption refers to the process by which a material, structure or object receives sound energy when sound waves are encountered, as opposed to reflecting the energy. Part of the absorbed sound energy is transformed into heat and part is transmitted through the absorbing body. The energy transformed into heat is said to have been 'lost'. Acoustic absorption is of particular interest in soundproofing.

Acoustic attenuation Acoustic attenuation is a measure of the energy loss of sound propagation in a particular media. Most media have viscosity and are therefore not ideal media. When sound propagates in such media, which has some viscosity there is always some energy loss by the viscosity.

Acoustic emission The detected energy that is generated when materials are deformed or break. For rolling element bearing analysis, it is the periodic energy generated by the over rolling of particles or flaws and detected by the display of the bearing flaw frequencies.

Acoustic impedance Acoustic impedance and specific acoustic impedance are measures of the resistance that a system presents to the acoustic flow, resulting an acoustic pressure applied to the system. The complex quotient is obtained when the sound pressure averaged over the surface is divided by the volume velocity through the surface. The real and imaginary components are known as

R. Padhye and R. Nayak (eds.), *Acoustic Textiles*, Textile Science and Clothing Technology, DOI 10.1007/978-981-10-1476-5

acoustic resistance and acoustic reactance, respectively. The SI unit of acoustic impedance is the Pascal second per cubic metre (Pa s/m^3) or the rayl per square metre (rayl/m^2), while that of specific acoustic impedance is the Pascal second per metre (Pa s/m) or the rayl.

Acoustic signal Acoustic signals are the noises that animals produce in response to a specific stimulus or situation, that have a specific meaning.

Acoustic signal processing It is the electronic manipulation of acoustic signals. Applications of acoustic signal processing include active noise control; design for hearing aids or cochlear implants; music information retrieval and perceptual coding.

Active noise control Active noise control (ANC), also known as noise cancellation is a method of reducing unwanted sound by the addition of a second sound designed to cancel the first. The noise-cancellation speaker emits a sound wave with the same amplitude but with opposite phase (also known as antiphase) to the original sound. The waves combine to form a new wave, in a process called interference, and effectively cancel each other out—an effect which is called destructive interference.

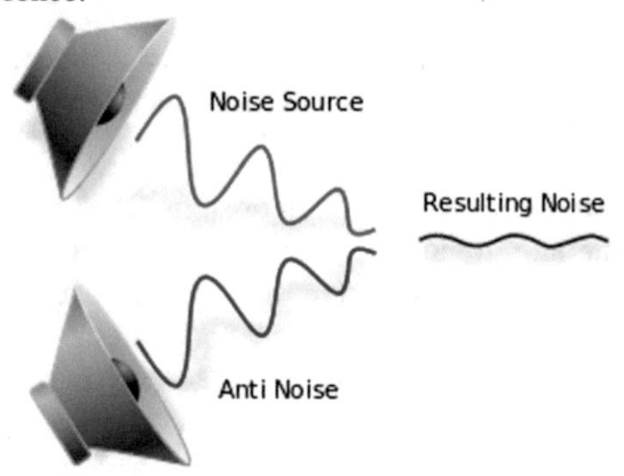

Airborne sound Sound that arrives at the point of interest, such as one side of a partition, by propagation through air. Airborne noise is transmitted in the form of sound waves by the progressive movement of mass particles (vibrations) at the speed of sound.

Air permeability Air permeability is a measure of how well the fabric or textile structure allows airflow through it under a differential pressure between the two surfaces (face and back). Air permeability is defined as the volume of air in millilitres, which is passed in one second through 1 cm^2 of a fabric at a pressure difference of 10 mm head of water.

Airflow resistance It is measured as a quotient of the air pressure difference across a specimen divided by the volume velocity of airflow through the specimen. The pressure difference and the volume velocity may be either steady or alternating.

Airflow resistivity For a homogeneous material, the quotient of its specific airflow resistance divided by its thickness.

Ambience The acoustic characteristics of a space with regard to reverberation. A room with a lot of reverb is said to be "live"; one without much reverb is said to be "dead".

Ambient noise The composite of airborne sound from many sources near and far associated with a given environment. No particular sound is singled out for interest.

AES Audio Engineering Society.

Amplifier It is an electronic device that can increase the power of a input signal. It does this by taking energy from a power supply and controlling the output to match the input signal shape but with larger amplitude.

Amplitude The amplitude of a wave is the distance from the centre line (or the still position) to the top of a crest or to the bottom of a trough. Amplitude is measured in metres (m) or centimetres (cm). The higher is the amplitude of a wave, the more energy is carried by it.

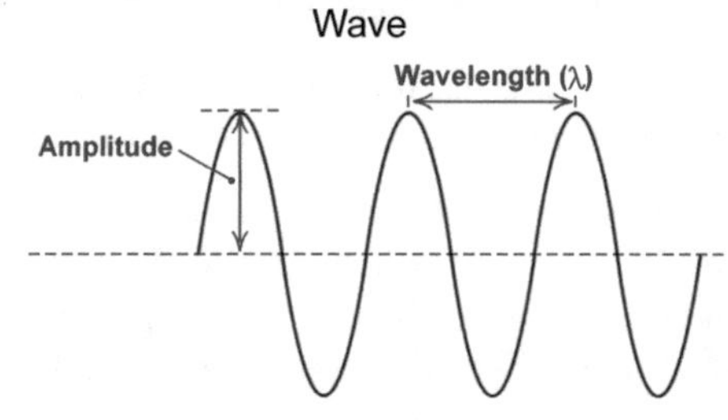

Analog An electrical signal whose frequency and level vary continuously in direct relationship to the original electrical or acoustical signal.

Antiaris toxicaria *Antiaris toxicaria* is a tree in the mulberry and fig family, Moraceae. It is the only species currently recognized in the genus Antiaris. It has a remarkably wide distribution in tropical regions, occurring in Australia, tropical Asia, tropical Africa, Indonesia, the Philippines, Tonga, and various other tropical islands.

Aqueous dispersion A liquid system in which very small solid particles are uniformly dispersed in water. These are two-phase liquid systems where one phase consists of finely divided particles of water insoluble, solid chemical products distributed throughout the second phase, which is water.

Areal density The areal density or surface density of a two-dimensional object is calculated as the mass per unit area. The SI derived unit is, kilogram or gram per square metre.

Attenuate To reduce the level of an electrical or acoustical signal.

Audible frequency range The range of sound frequencies normally heard by the human ear. The audible range spans from 20 to 20,000 Hz.

Audible sound Vibrations transmitted through an elastic solid or a liquid or gas, with frequencies in the approximate range of 20–20,000 Hz, capable of being detected by human organs of hearing.

Background noise Noise from all sources unrelated to a particular sound that is the object of interest. Background noise may include airborne, structure-borne and instrument noise. In acoustical engineering, background noise or ambient noise is any sound other than the sound being monitored (primary sound). Background noise is a form of noise pollution or interference. Background noise is an important concept in setting noise regulations.

Baffle A baffle or sound baffle is a movable barrier used to achieve separation of signals from different sources. A sound baffle is also used to reduce the strength (level) of airborne sound. Sound baffles are fundamental tools for noise mitigation, the practice of minimizing noise pollution or reverberation.

Band width The total frequency range of any system. Bandwidth is usually expressed in a range, such as 20–20,000 Hz.

Bicomponent fibres Bicomponent fibres are produced by extruding two polymers from the same spinneret with both polymers contained within the same filament.

Blowing agent A blowing agent is a substance which is capable of producing a cellular structure via a foaming process in a variety of materials that undergo hardening or phase transition, such as polymers, plastics and metals. They are typically applied when the blown material is in a liquid stage.

Braiding A braid (also referred to as a plait) is a complex textile structure formed by interlacing three or more strands of flexible material such as yarns. Compared with weaving, which usually involves two separate, perpendicular groups of strands (warp and weft), a braided structure is usually long and narrow, with each component strand functionally equivalent in zigzagging forward through the overlapping mass of the others.

Bursting strength It is the pressure at which a film or sheet of paper or plastic will burst. It is used as a measure of resistance to rupture, which depends largely on the tensile strength and extensibility of the material.

Butyl rubber Butyl rubber is a synthetic rubber produced by the polymerization of isobutene with a small amount of isoprene. It is impermeable to air and has several applications ranging from inner tubes for tyres to adhesives to sealants.

Carbonization It is the process of conversion of an organic substance into carbon or a carbon-containing residue through pyrolysis or destructive distillation. It is often used in organic chemistry with reference to the generation of coal gas and coal tar from raw coal.

Carpet A carpet is a textile floor covering typically consisting of an upper layer of pile attached to a backing. Carpets are used in automotive and interior design for sound absorption. Carpets can be of different structures such as woven, needle felt, knotted, tufted or any other modification prepared from nylon, polyester, wool and its blends, acrylic and polypropylene.

Coated or laminated textiles Coating and lamination techniques are used to impart functionality to woven or knitted fabrics, which are not necessarily those naturally assumed by textile fabrics. Having widespread application across a range of technical textiles sectors, they increase functionality and durability as well as value. They can include: waterproofness, increased abrasion, stain, flame and UV resistance, retro-Reflection, or anti-microbial or Phase Change Materials.

Compression In audio, compression means to reduce the dynamic range of a signal. Compression may be intentional or one of the effects of a system that is driven to overload. It is also the portion of a sound wave in which molecules are pushed together, forming a region with higher-than-normal atmospheric pressure. For fibrous structures, compression indicates the application of load to reduce the size. Compressive strength is the capacity of a structure to withstand the loads tending to reduce size, as opposed to tensile strength, which withstands loads tending to elongate. The compression results in decrease of the thickness and the porosity.

Compression moulding Compression moulding is a method of moulding in which the moulding material, generally preheated, is first placed in an open, heated mould cavity. The mould is closed with a top force or plug member, pressure is applied to force the material into contact with all mould areas, while heat and pressure are maintained until the moulding material has cured. The process employs thermosetting resins in a partially cured stage, either in the form of granules, putty-like masses, or preforms.

Computational fluid dynamics Computational fluid dynamics, usually abbreviated as CFD, is a branch of fluid mechanics that uses numerical analysis and algorithms to solve and analyze problems that involve fluid flows.

Creep Creep may be defined as a time-dependent deformation at elevated temperature and constant stress. It is the tendency of a nonwoven fabric to slowly move or deform permanently under the influence of stresses. It follows, then, that a failure from such a condition is referred to as a creep failure or occasionally, a stress rupture.

Critical band In human hearing, only those frequency components within a narrow band, called the critical band, will mask a given tone. Critical bandwidth varies with frequency but is usually between 1/6 and 1/3 octaves.

Critical damping The smallest amount of damping required to return a system to its equilibrium condition without oscillating.

Critical distance The distance from a sound source at which direct sound and reverberant sound are at the same level.

Critical frequency The frequency below which standing waves cause significant room modes.

Cycles per second The frequency of an electrical signal or sound wave. Measured in Hertz (Hz).

Damping Damping is an influence within or upon an oscillatory system that has the effect of reducing, restricting or preventing its oscillations. In physical systems, damping is produced by processes that dissipate the energy stored in the oscillation.

Damping factor or damping ratio The ratio of actual damping in a system to its critical damping.

Decibel (dB) It is a unit used to measure the intensity of a sound or the power level of an electrical signal by comparing it with a given level on a logarithmic scale. The term used to identify ten times the common logarithm of the ratio of two like quantities proportional to power or energy.

Diffraction A change in the direction of propagation of sound energy in the neighbourhood of a boundary discontinuity, such as the edge of a reflective or absorptive surface or obstacle. Diffraction occurs in water waves, sound waves and light waves, but the amount of diffraction depends on the size of the obstacle or opening in relation to the wavelength of the wave.

Diffuse field An environment in which the sound pressure level is the same at all locations and the flow of sound energy is equally probable in all directions.

Digital A numerical representation of an analog signal. Pertaining to the application of digital techniques to common tasks.

Divergence The spreading of sound waves which, in a free field, causes sound pressure levels in the far field of a source to decrease with increasing distance from the source.

Doppler Effect Doppler Effect or the Doppler Shift is the change in frequency of a wave (or other periodic event) for an observer moving relative to its source. It was established by the Austrian physicist Christian Doppler, in 1842 in Prague. It is commonly heard when a vehicle sounding a siren or horn approaches, passes and recedes from an observer.

DREF yarns The yarns produced by friction spinning or DREF spinning, which is suitable for spinning coarse yarns and technical core-wrapped yarns. DREF yarns are bulky, with low tensile strength making them suitable for blankets and mop yarns, they can be spun from asbestos, carbon fibres and make filtres for water systems.

Drape The way in which a garment or fabric hangs.

Duct liner Duct liner is used primarily as acoustical liner in heating, ventilation, and air conditioning (HVAC) sheet metal ducts to absorb unwanted crosstalk, equipment and air rush noise.

Ear mufflers Ear mufflers or earmuffs are objects designed to cover a person's ears for protection such as sound or for warmth.

Echo A delayed return of sound that is perceived by the ear as a discrete sound image.

Ecolabelling It is an environmental performance certification and labelling system that is practised around the world. An ecolabel identifies products or services proven environmentally preferable overall, within a specific product or service category.

Elasticity It is the ability of a material to resist a distorting force or stress and to return to its original size and shape when the stress is removed.

Elbow silencers Elbow silencers provide a solution to undesirable noise in short runs of ductwork. They are used in places where straight silencers cannot be used.

Electrospinning Electrospinning is a fibre production technique which uses electrostatic force to draw charged threads of polymer solutions or polymer melts up to fibre diameters in the order of some ten nanometres. Electrospinning shares characteristics of both electrospraying and conventional solution dry spinning of fibres.

Environmental acoustics It focuses on controlling and restraining the impact of noise on the environment, caused by sources such as companies and leisure facilities.

Environmental noise It is the noise pollution from external sources, caused by transportation, industrial and recreational activities.

Epoxy resin Epoxy resins are low molecular weight pre-polymers or higher molecular weight polymers, which normally contain at least two epoxide groups. The raw materials for epoxy resin production are today largely petroleum derived, although some plant-derived sources are now becoming commercially available.

Estabragh fibres Estabragh (*Asclepias procerais*) fibres are natural, hollow fibres, which are expected to provide excellent insulation properties due to their individual characteristics. The mechanical behaviour of these fibres plays a major role in mechanical processing.

Extension How extended a range of frequencies the device can reproduce accurately. Bass extension refers to how low a frequency tone will the system reproduce, high frequency extension refers to how high in frequency will the system play.

Fancy twisting machine This machine is used for the production of modern fancy yarns for clothing and upholstery fabrics. This machine can produce fancy yarns such as boucle, loops, braid, spiral, knots, wave, flame, thick and thin.

Far field That part of the sound field in which sound pressure decreases inversely with distance from the source. This corresponds to a reduction of approximately 6 dB in level for each doubling distance.

Fibre fineness Fibre fineness is one the most important fibre characteristics, which determines how many fibres are present in the cross-section of a yarn of given thickness.

Fibrous material A fibrous material is produced from fibres.

Ficus brachypoda *Ficus brachypoda* is a tree in the family Moraceae native to northern Australia. It is a banyan of the genus Ficus which contains around 750 species worldwide in warm climates, including the edible fig (*Ficus carica*).

Ficus natalensis *Ficus natalensis* is a tree in the family Moraceae. It is commonly known as the Natal fig or "Mutuba" to locals. These trees are distributed from north-eastern South Africa to Uganda and Kenya.

Finite element analysis Finite element analysis is a numerical technique for finding approximate solutions to boundary value problems for partial differential equations. It is also referred to as finite element method (FEM). FEM subdivides a large problem into smaller, simpler and parts called finite elements. FEM can be used to predict how a product reacts to real-world forces, vibration, heat, fluid flow and other physical effects.

Flame retardant Flame retardants are the materials used for reducing the devastating impact of fires on people, property and the environment. They are added to or treat potentially flammable materials, including textiles and plastics.

Fidelity As applied to sound quality, the faithfulness to the original.

Fourier transform The mathematically rigorous operation which transforms from the time domain to the frequency domain and vice versa.

Frequency The number of waves that pass through a fixed place in a given amount of time. Alternately, frequency is the measure of the rapidity of alterations of a periodic signal, expressed in cycles per second or Hz. For example, if the time taken for a wave to pass a place is 0.5 s, the frequency is 2 cycles per second.

Glycerol It is a colourless, odourless, viscous liquid that is sweet-tasting and non-toxic. It is widely used in the food industry as a sweetener and humectant and in pharmaceutical formulations.

Hand lay-up technique Hand lay-up technique is the simplest and oldest open moulding method of the composite fabrication processes. It is a low volume, labour-intensive method suited especially for large components, such as boat hulls.

Hearing loss Hearing loss, also known as hearing impairment, is a partial or total inability to hear. A deaf person has little to no hearing. Hearing loss may occur in one or both ears. In children, hearing problems can affect the ability to learn language and in adults it can cause work related difficulties.

Hertz (Hz) Hertz is the unit of frequency, abbreviated Hz, which is the same as cycles per second. The name was derived in the honour of Heinrich Hertz, an early German investigator of radio wave transmission.

Helmholtz resonator Helmholtz resonance is the phenomenon of air resonance in a cavity, such as when one blows across the top of an empty bottle. The name is derived from a device created by Hermann von Helmholtz in the 1850s. The resonator was use by Helmholtz to identify various frequencies or musical pitches present in music and other complex sound.

Highway noise Highway or roadway noise is the total sound energy emanating from moving motor vehicles. It consists of friction of tyre with road, engine/transmission, aerodynamic and braking elements.

Hydroentanglement Hydroentanglement is a bonding process for wet or dry fibrous webs made by carding, air-laying or wet-laying, resulting in a nonwoven fabric. It uses fine, high pressure jets of water which penetrate the web, hit the conveyor belt and bounce back causing the fibres to entangle. Hydroentanglement is sometimes known as spunlacing.

Hydrophobic In chemistry, hydrophobicity is the physical property of a molecule (known as a hydrophobe) that is seemingly repelled from a mass of water. Hydrophobic molecules tend to be non-polar and, thus, prefer other neutral molecules and non-polar solvents.

Hypersound Sound waves with frequencies ranging from 10^9 to 10^{12} or 10^{13} Hz; the high frequency portion of the spectrum of elastic waves. The physical nature of hypersound is not different from that of ultrasound, whose frequency ranges from 2×10^4 to 10^9 Hz.

Hysteresis Non-uniqueness in the relationship between two variables as a parameter increases or decreases.

Image J Image J is an open source image processing program designed for scientific multidimensional images. Image J can display, edit, analyze, process, save and print 8-bit colour and grayscale, 16-bit integer and 32-bit floating point images.

Impact sound transmission test Standard test method used for laboratory measurement of impact sound transmission through floor-ceiling.

Impedance The opposition to the flow of electric or acoustic energy measured in ohms.

Impedance tube The test set-up consisting of standing wave tube (also called an impedance tube) method is used to measure the sound absorption coefficient and impedance of various materials.

Impulse A very short, transient, electric or acoustic signal.

Impulse response Sound pressure versus time measurement showing how a device or room responds to an impulse.

In phase Two periodic waves reaching peaks and going through zero at the same instant are said to be "in phase."

Infrasound Infrasound is sound below 20 Hz, lower than humans can perceive. Infrasound is produced by elephants, in particular, that travel through solid ground and are sensed by other herds using their feet, although they may be separated by hundreds of kilometres.

Interference The combining of two or more signals results in an interaction called interference. This may be constructive or destructive. Another use of the term is to refer to undesired signals.

Intensity Acoustic intensity is sound energy flux per unit area. The average rate of sound energy transmitted through a unit area normal to the direction of sound transmission.

Kawabata Evaluation System (KES) KES is used to measure the low-stress mechanical properties of fabrics. The system was developed by a team led by Professor Kawabata in the department of polymer chemistry, Kyoto University (Japan). KES consists of four different modules on which a total of six tests can be performed such as, tensile and shear (Module 1); bending (Module 2); compression (Module 3); and surface friction and surface roughness (Module 4).

kHz 1000 Hz.

Knitting Knitting is a process by which yarn is manipulated to create a fabric by formation of loops. Knitting creates multiple loops of yarn, called stitches, in a line or tube. Knitting has multiple active stitches on the needle at one time. Knitting can be two types, weft knitting and warp knitting. In the weft knitting, the entire fabric may be produced from a single yarn, by adding stitches to each wale in turn, moving across the fabric; whereas in warp knitting, one yarn is required for every wale. Since a piece of knitted fabric may have hundreds of wales, warp knitting is typically done by machines, whereas weft knitting is done by both hand and machine. The common knit structures are plain, rib and interlock. A text book on knitting can provide the difference between these structures.

Knitted fabric Knitted fabric is a textile material produced by knitting. Its properties are distinct from woven fabric in that it is more flexible and can be more readily constructed into smaller pieces making it ideal for socks and hats.

Kundt's tube Kundt's tube is an acoustical experimental apparatus invented in 1866 by German physicist, August Kundt for the measurement of speed of sound in a gas or a solid rod. It is used today only for demonstrating standing waves and acoustical forces.

Linear A device or circuit with a linear characteristic means that a signal passing through it is not distorted.

Lignocellulosic fibres Lignocellulosic materials are any of several closely related substances constituting the essential part of woody cell walls of plants and consisting of cellulose intimately associated with lignin. Lignocellulosic fibres include jute, sisal, pineapple and curauá.

Longitudinal wave Longitudinal wave is the wave in which the displacement of the medium occurs in the same direction as or the opposite direction to, the direction of travel of the wave.

Loudspeaker An electro-acoustical transducer that changes electrical energy to acoustical energy.

Loudness Loudness is the characteristic of a sound that is primarily a psychological correlate of physical strength (amplitude). Loudness can be defined as that attribute of auditory sensation in terms of which sounds can be ordered on a scale extending from quiet to loud.

Mechanical properties The mechanical properties of a material describe how it will react to physical forces. Mechanical properties occur as a result of the physical properties inherent to each material and determined through a series of standardized mechanical tests such as strength, ductility, hardness, impact resistance, and fracture toughness.

Mechanical wave A mechanical wave is a wave which is an oscillation of matter and therefore transfers energy through a medium.

Medium An intervening substance or agency for transmitting or producing an effect. For example, air is a medium for sound.

Meltblowing Meltblowing (MB) is a process for producing fibrous webs or articles directly from polymers or resins using high-velocity air or another appropriate force to attenuate the filaments. This process is unique because it is used almost exclusively to produce microfibres rather than fibres the size of normal textile fibres.

Metric Sabin The unit of measure of sound absorption in the MKS (metre-kilogram-second) system of units.

Microfibre Microfibre is a synthetic fibre finer than one denier or decitex/thread. The most common types of microfibres are made from polyesters, polyamides. Microfibre is used to make mats, knits, and weaves for apparel, upholstery, industrial filtres and cleaning products.

Microphone An acoustical-electrical transducer by which sound waves in air may be converted to electrical signals, which may then be amplified, transmitted, or recorded.

MPa (Megapascal) It is a pressure measurement unit. 1 MPa = 1 Newton/millimetre^2 (N/mm^2).

Nanomaterial A material having particles or constituents of nanoscale dimensions or one that is produced by nanotechnology.

Natural fibres Natural fibres are the fibres that are obtained from plant, animal and mineral sources. Those from plant sources include cotton, flax, hemp, sisal, jute, kenaf and coconut. Fibres from animal sources include silk, wool and mohair. Those from mineral sources include asbestos, glass and metal fibres.

Near field Locations close to the sound source between the source and the far field. The near field is typically characterized by large sound pressure level variations with small changes in measurement position from the source.

Needle felt carpets These are produced by intermingling and felting individual synthetic fibres using barbed and forked needles forming an extremely durable carpet. These carpets are normally found in commercial settings such as hotels and restaurants where there is frequent traffic.

Needle punching Needle punching is the oldest method of producing nonwoven products. The needle punching system is used to bond dry laid and spun laid webs. The needle punched fabrics are produced when barbed needles are pushed through a fibrous web forcing some fibres through the web, where they remain when the needles are withdrawn.

Noise Noise is the sound that is not wanted by the perceiver, as it is unpleasant, loud or interferes with hearing. High levels of noise have negative impacts from mere inconvenience to hearing damage on humans. They can also cause environmental damage as animals are affected as well.

Noise barrier Noise or acoustical barrier is an exterior structure designed to protect inhabitants of sensitive land use areas from noise pollution. Noise barriers are the most effective method of mitigating roadway, railway and industrial noise sources, other than cessation of the source activity or the use of source controls.

Noise control Noise control or noise mitigation is a set of strategies to reduce noise pollution or to reduce the impact of, whether outdoors or indoors.

Noise pollution Noise pollution is the excessive noise that may harm the activity or imbalance the life of human or animal. Sound becomes unwanted or noise when it either interferes with normal activities such as sleeping, conversation or disrupts or diminishes one's quality of life.

Noise reduction (NR) The difference in sound pressure level between any two points along the path of sound propagation. As an example, noise reduction is the term used to describe the difference in sound pressure levels between the inside and outside of an enclosure.

Noise reduction coefficient (NRC) The Noise Reduction Coefficient (commonly known as NRC) is a scalar representation of the amount of sound energy absorbed upon striking a particular surface. An NRC of 0 indicates perfect reflection whereas an NRC of 1 indicates perfect absorption.

Noise regulation Noise regulation includes laws or guidelines relating to sound transmission established by national or state or provincial and municipal levels of government. A noise regulation restricts the amount of noise, the duration of noise and the source of noise. It usually places restrictions for certain times of the day or night on the amount of noise.

Nonwoven composite These are the composite materials comprising multi-fibrous layers. These composites have attracted considerable attention owing to their excellent thermal properties, high porosity and cost competitiveness.

Nonwoven fabric It is a fabric-like material made from long fibres, bonded together by chemical, mechanical, thermal or solvent treatment. The term is used in the textile manufacturing to denote fabrics, such as felt, which are neither woven nor knitted. In recent years, nonwovens have become an alternative to polyurethane foam.

Normal incidence sound absorption Normal incidence implies the sound application to the specimen in a perpendicular direction. For a surface, it the measure of the fraction of the perpendicularly incident sound absorbed or otherwise not reflected.

Octave An octave is a doubling or halving of frequency. 20–40 Hz is often considered the bottom octave.

Octave bands Frequency ranges in which the upper limit of each band is twice the lower limit. For example, an octave filtre with a centre frequency of 1 kHz has a lower frequency of 707 Hz and an upper frequency of 1.414 kHz. Octave bands are identified by their geometric mean frequency, or centre frequency. Fractional-octave analysis is a technique for analyzing audio and acoustic signals. Fractional-octave analyzes (especially 1/3 and 1/12 octave) exhibits characteristics analogous to the response of the human ear.

One-third octave bands Frequency ranges where each octave is divided into one-third octaves with the upper frequency limit being 2 times the lower frequency. It is identified by the geometric mean frequency of each band.

Oxidation Oxidation is the loss of electrons or an increase in oxidation state by a molecule, atom, or ion.

Panel absorbers Acoustic panel absorbers are functional and aesthetically pleasing perforated panels used to control reverberant noise in factories, waste water treatment facilities, pump rooms and other environmental applications.

Particle size distribution The particle-size distribution of a powder or granular material dispersed in fluid, is a list of values or a mathematical function that defines the relative amount (typically by mass) of particles present according to size.

Particle velocity Particle velocity is the velocity of a particle (real or imaginary) in a medium as it transmits a wave. The SI unit of particle velocity is the metre per second (m/s). In the case of a sound wave through a medium of a fluid like air, particle velocity would be the physical speed of a parcel of fluid as it moves back and forth in the direction the sound wave is travelling as it passes.

Passive absorber Passive absorbers aim to suppress the sound by modifying the environment close to the source. Since no input power is required, passive absorbers are often cheaper than active absorbers. However, the performance is limited to mid and high frequencies. Active control works well for low frequencies. Hence, the combination of two methods may be utilized for broadband noise reduction. Active sound absorber dissipates sound energy as heat.

Peak The maximum positive or negative dynamic excursion from zero (for an AC coupled signal or a sound wave) or from the offset level (for a DC coupled) of any time waveform.

Peak-to-peak The amplitude difference between the most positive and most negative value in the time waveform.

Period A signal that repeats the same pattern over time is called periodic and the period is defined as the length of time encompassed by one cycle, or repetition. The period of a periodic waveform is the inverse of its fundamental frequency.

Periodic A signal is periodic if it repeats the same pattern over time. The spectrum of a periodic signal always contains a series of harmonics.

Phase Phase is the measure of progression of a periodic wave. Phase identifies the position at any instant which a periodic wave occupies in its cycle. It can also be described as the time relationship between two signals.

Phase shift The time or angular difference between two signals.

Phase (time lag or lead) The difference in time between two events such as the zero crossing of two waveforms, or the time between a reference and the peak of a waveform. The phase is expressed in degrees as the time between two events divided by the period (also a time), times 360°.

Phon The unit of loudness level of a tone.

Pile height It is the thickness of a rug or carpet measured from the surface of a rug or carpet to its backing.

Pitch The quality of a sound governed by the rate of vibrations producing it; the degree of highness or lowness of a tone.

Polylactic acid Polylactic acid (PLA) or polylactide is a biodegradable thermoplastic aliphatic polyester derived from renewable resources, such as corn starch (in the United States and Canada), tapioca roots, chips or starch (mostly in Asia), or the sugarcane (in the rest of the world).

Polymer A polymer is a large molecule or macromolecule composed of many repeated units. Because of their wide range of properties, both synthetic and natural polymers play an essential and ubiquitous role in everyday life.

Polymer composite Polymer composites combine a resin system and reinforcing fibres. The properties of the resulting composite material will combine some of the properties of the resin on its own with that of the fibres on their own. Exceptional properties can be obtained by polymer composites when the resin systems are combined with reinforcing fibres such as glass, carbon and aramid.

Polyurethane Polyurethane (PU) is a polymer composed of organic units joined by carbamate (urethane) links. While most polyurethanes are thermosetting polymers that do not melt when heated, thermoplastic polyurethanes are also available. PU polymers are most commonly formed by the reaction of a di- or poly-isocyanate with a polyol. Both the isocyanates and polyols used to make polyurethanes contain on an average two or more functional groups per molecule.

Polyvinyl alcohol Polyvinyl alcohol (PVA) is a colourless and water-soluble synthetic resin employed principally in the treating of textiles and paper. PVA is unique among polymers in that it is not built up in polymerization reactions from single-unit precursor molecules known as monomers. Instead, PVA is made by dissolving another polymer, polyvinyl acetate (PVAc), in an alcohol such as methanol and treating it with an alkaline catalyst such as sodium hydroxide.

Porosity Porosity is a measure of the void or empty spaces in a material, and is a fraction of the volume of voids over the total volume. It ranges from 0 to 1 or as a percentage from 0 to 100 %. Porosity plays a very important role in sound absorption.

Porous absorber A porous absorber is any fibrous or porous material such as textiles, fleece, foams that can absorb sound energy as they damp the oscillation of the air particles by friction. These absorbers are the most effective in slowing down air particles with a high sound velocity.

Precursor In chemistry, a precursor is a compound that participates in a chemical reaction that produces another compound. In biochemistry, it often refers more specifically to a chemical compound preceding another in a metabolic pathway, such as a protein precursor.

Pressure wave A wave (such as a sound wave) in which the propagated disturbance is a variation of pressure in a material medium such as air.

Radiation Radiation is the energy that comes from a source and travels through some material or through space. Light, heat and sound are types of radiation.

Reactive absorber A sound absorber, such as the Helmholtz resonator which involves the effects of mass and compliance as well as resistance.

Receiving room In architectural acoustical measurements, the room in which the sound transmitted from the source room is measured.

Release agent The primary purpose of the release agent is to prevent the resin from sticking to the mould.

Recycling Recycling is the process of converting waste materials into reusable products. Recycling can prevent waste of potentially useful materials, reduce the consumption of fresh raw materials, energy usage, air pollution and water pollution by decreasing the need for conventional waste disposal and lowering greenhouse gas emissions compared to plastic production.

Recycled material Materials produced by recycling process.

Reflection Reflection is the change in direction of a wavefront at an interface between two different media so that the wavefront returns into the medium from which it originated. Common examples include the reflection of light, sound and water waves. The law of reflection says that for specular reflection the angle at which the wave is incident on the surface equals the angle at which it is reflected. In acoustics, reflection causes echoes and is used in sonar.

Refraction Refraction is the change in direction of propagation of a sound wave due to a change in its transmission medium. The phenomenon is explained by the conservation of energy and the conservation of momentum. Due to the change of medium, the phase velocity of the wave is changed but its frequency remains constant.

Resistance That quality of electrical or acoustical circuits that result in dissipation of energy through heat.

Resonance It is a natural periodicity or the reinforcement associated with this periodicity. In sound applications, a resonant frequency is a natural frequency of vibration determined by the physical parameters of the vibrating object.

Resonant absorbers Resonant absorbers are the most powerful sound absorbers for low frequency sound absorption. These absorbers consist of a mechanical or acoustical oscillation system. Resonant absorbers work by vibrating at these low frequencies and turning the sound energy into heat.

Resonant frequency Any system has a resonance at some particular frequency. At that frequency, even a slight amount of energy can cause the system to vibrate. A stretched piano string, when plucked, will vibrate for a while at a certain fundamental frequency. Plucked again, it will vibrate at that same frequency. This is its natural or resonant frequency. While this is the basis of musical instruments, it is undesirable in music-reproducing instruments like audio equipment.

Reverberant sound field The sound in an enclosed or partially enclosed space that has been reflected repeatedly or continuously from the boundaries.

Reverberation Reverberation, in acoustics and psychoacoustics, is the persistence of sound after a sound is produced. A reverberation or reverb is created when a sound is reflected causing a large number of reflections to build up and then decay as the sound is absorbed by the surfaces of objects in the space, which could include furniture, people, and air. This is most noticeable when the sound source stops but the reflections continue, decreasing in amplitude, until they reach zero amplitude.

Reverberation room A reverberation chamber or room is a room designed to create a diffuse or random incidence sound field (i.e. one with a uniform distribution of acoustic energy and random direction of sound incidence over a short time period). Reverberation rooms tend to be large rooms (the resulting sound field becomes more diffused with increased path length) and have very hard exposed surfaces. The change of impedance (compared to the air) of these surfaces present to incident sound is so large that virtually all of the acoustic energy that hits a surface is reflected back into the room. Reverberation rooms are used in acoustics as well as in electrodynamics, such as for measurement microphone calibration, measurement of the sound power of a source, and measurement of the absorption coefficient of a material.

Reverberation time The time taken for a sound to drop by 60 dB is known as the reverberation time. RT60 is the time required for reflections of a direct sound to decay 60 dB. Reverberation time is frequently stated as a single value, if measured as a wide band signal (20–20 kHz). However, being frequency dependent, it can be more precisely described in terms of frequency bands (one octave, 1/3 octave, 1/6 octave, etc.). The reverberation time measured in narrow bands will differ depending on the frequency band being measured. For precision, it is important to know what ranges of frequencies are being described by a reverberation time measurement.

Root mean square In statistics, the root mean square (abbreviated RMS or rms), also known as the quadratic mean, is defined as the square root of the arithmetic mean of the squares of a set of numbers.

RT60 RT60 is an acoustical measurement used to calculate the reverb time decay. RT60 is in reality the measurement of time it takes a given audio signal to fall by 60 dB (decibels).

Sabin The unit of measure of sound absorption in the inch-pound system.

Sabine's formula Wallace Clement Sabine (1868–1919), an American physicist, is considered the father of architectural acoustics. He is the originator of the Sabine reverberation equation for determining the quality of sound in a lecture hall, auditorium, or any other venue where people speak or play music. Sabine derived an expression for the duration, T, of the residual sound to decay below the audible intensity, starting from a 1,000,000 times higher initial intensity as in the following equation,

$$T = 0.161\frac{V}{A}$$

where V is the room volume in cubic metres, and A is the absorption coefficient.

Scattering Scattering is a general physical process where some forms of radiation, such as light, sound or moving particles are forced to deviate from a straight trajectory by one or more paths due to localized non-uniformities in the medium through which they pass.

Scattering coefficient Scattering coefficient expresses the attenuation caused by scattering, e.g., of radiant or acoustic energy, during its passage through a medium. The scattering coefficient is usually expressed in units of reciprocal distance.

Shear thickening fluid A shear thickening fluid is a material whose viscosity increases with the rate of shear strain. Such a shear thickening fluid, also known by STF, is an example of a non-Newtonian fluid. This behaviour is only one type of deviation from Newton's Law and it is controlled by such factors as particle size, shape and distribution.

Shoddy An inferior quality yarn or fabric made from the shredded fibre of waste woollen cloth or clippings.

Silencer Silencer may refer to a device for reducing the amount of noise emitted by the exhaust of an internal combustion engine.

Signal-to-noise (SN) Ratio Signal-to-noise ratio (abbreviated SNR or SN) is a measure used in science and engineering that compares the level of a desired signal to the level of background noise. It is defined as the ratio of signal power to the noise power, often expressed in decibels. A ratio higher than 1:1 (>0 dB) indicates more signal than the noise.

Sine wave A periodic wave related to simple harmonic motion.

Soft room A room with highly sound absorptive surfaces.

Sone The unit of measurement for subjective loudness.

Sound Sound is vibrational disturbance, exciting hearing mechanisms, transmitted in a predictable manner determined by the medium through which it propagates. To be audible, the sound must fall within the frequency range 20–20,000 Hz.

Sound absorption See acoustic absorption.

Sound absorption class The acoustic materials are classified into five classes, A to E depending on their sound absorption capacity, with A having the highest capacity to absorb sound.

Sound absorption coefficient The fraction of sound energy absorbed by a material. It is expressed as a value between 0 (no absorption, total reflection) and 1.0 (perfect absorption, no reflection).

Sound attenuation The reduction of the intensity of sound as it travels from the source to a receiving location. See Acoustic attenuation.

Sound energy, *E* Sound energy is a form of energy that is associated with vibrations of matter. It is a type of mechanical wave which requires an object such as air and water to travel through. The SI unit of sound energy is the joule (J).

Sound energy density, *D* The quotient obtained when the sound energy in a region is divided by the volume of the region. The sound energy density at a point is the limit of that quotient as the volume that contains the point approaches zero.

Sound insulation The capacity of a structure to prevent sound from reaching a receiving location. Sound insulation measures the ability of building elements or structures to reduce sound transmission. Sound energy is not necessarily absorbed; impedance mismatch or reflection back toward the source is often the principal mechanism.

Sound intensity, *I* Sound intensity or acoustic intensity is the sound power per unit area. The SI unit of sound intensity is watt per square metre (W/m^2). The usual context is the noise measurement of sound intensity in the air at a listener's location as a sound energy quantity. The quotient obtained when the average rate of energy flow in a specified direction and sense is divided by the area, perpendicular to that direction, through or toward which it flows. The intensity at a point is the limit of that quotient, the area that includes the point approaches zero.

Sound isolation The degree of acoustical separation between two locations, especially adjacent rooms.

Sound level Of airborne sound, a sound pressure level obtained using a signal to which a standard frequency weighting has been applied.

Sound power, W Sound power or acoustic power is the rate at which sound energy is emitted, reflected, transmitted or received, per unit time. The SI unit of sound power is the watt (W). It is the power of the sound force on a surface of the medium of propagation of the sound wave.

Sound power level Of airborne sound, ten times the common logarithm of the ratio of the sound power under consideration of the standard reference power of 1 pW. The quantity so obtained is expressed in decibels.

Sound pressure Sound pressure or acoustic pressure is the local pressure deviation from the ambient (average, or equilibrium) atmospheric pressure, caused by a sound wave. In analogy with alternating voltage its magnitude can be expressed in several ways, such as instantaneous sound pressure or peak sound pressure, but the unqualified term means root-mean-square sound pressure.

Sound pressure Level (SPL) Given in decibels (dB) is an expression of loudness or volume. A 10 dB increase in SPL represents a doubling in volume. Live orchestral music reaches brief peaks in the 105 dB range and live rock easily goes over 120 dB.

Soundproofing It is the means of reducing the sound pressure with respect to a specified sound source and receiver. Soundproofing can suppress unwanted indirect sound waves such as reflections that cause echoes and resonances that cause reverberation. Soundproofing can reduce the transmission of unwanted direct sound waves from the source to an involuntary listener through the use of distance and intervening objects in the sound path. Approaches such as increasing the distance between the source and receiver, using noise barriers to reflect or absorb the energy of the sound waves, using damping structures such as sound baffles or using active antinoise sound generators can help in sound proofing.

Sound propagation Sound is a sequence of waves of pressure, which propagates through compressible media such as air or water. Although sound can propagate through solids as well, it needs additional modes of propagation.

Sound receiver One or more observation points at which sound is evaluated or measured. The effect of sound on an individual receiver is usually evaluated by measurements near the ear or close to the body.

Sound source The point from which sound waves are generated.

Sound spectrum A sound spectrum displays all the frequencies present in a sound. A sound spectrum is a representation of a sound—usually a short sample of a sound—in terms of the amount of vibration at each individual frequency. It is usually presented as a graph of either power or pressure as a function of frequency.

Sound transmission Sound or acoustic transmission in building design refers to a number of processes by which sound can be transferred from one part of a building to another.

Sound transmission class, STC Sound Transmission Class is an integer rating of how well a building partition attenuates airborne sound. The STC is calculated in accordance with Classification E413 using values of sound transmission loss. The STC number is derived from sound attenuation values tested at sixteen standard frequencies from 125 to 4000 Hz. It provides an estimate of the performance of a partition in certain common sound insulation problems.

Sound transmission loss (TL) Transmission loss describes the accumulated decrease in intensity of a sound energy as the wave propagates outwards from a source. TL is very important in the industry of acoustic devices such as mufflers and sonars. The unit of TL is dB. TL is defined by the following equation,

$$\mathrm{TL} = 10\log_{10}\left|\frac{W_i}{W_t}\right|$$

where W_i is the power of incident wave coming towards a defined area (or structure) and W_t is the power of transmitted wave going away from the defined area (or structure).

Sound wave Sound waves are similar to the waves in the water, characterized by wavelength and amplitude. When vibration sound disturbs surrounding particles in the air or liquid, those particles displace the particles next to them. This particle movement goes on continuously in outward direction to form a wave pattern. The wave carries the sound energy through the medium and becomes less intense as it moves away from the source.

Source room In architectural acoustical measurements, the room that contains the noise source or sources.

Specific airflow resistance, *r* The product of the airflow resistance of a specimen and its area. This is equivalent to the quotient of the air pressure difference across the specimen divided by the linear velocity, measured outside the specimen, of airflow through the specimen.

Spectrum The distribution of the energy of a signal with frequency.

Speed of sound The speed of sound is the distance travelled by sound per unit time by a sound wave propagating through an elastic medium. In dry air at 20 °C (68 °F), the speed of sound is 343.2 m/s.

Spunbonding A process for forming nonwoven fabrics, usually of limited durability, by bonding continuous-filament synthetic fibres immediately after extrusion.

Standing wave A resonance condition in an enclosed space in which sound waves travelling in one direction interact with those travelling in the opposite direction, resulting in a stable condition.

Steady-state A condition devoid of transient effects.

Structure-borne noise Structure-borne noise is transmitted through solid structures, such as steel, wood, concrete, stone etc., which result in unwanted radiated sound. This includes for example impact sound and part of the noise generated by the technical machinery installed in a building.

Superposition Many sound waves may transverse the same point in space, the air molecules responding to the vector sum of the demands of the different waves.

Synthetic fibres Synthetic fibres are the fibres that are created by extruding fibre forming materials through spinnerets into air and water, forming a thread. Synthetic fibres are made from synthesized polymers or small molecules. The compounds that are used to make the synthetic fibres come from petroleum based raw materials. These materials are polymerized into a long, linear chemical that bond two adjacent carbon atoms. Differing chemical compounds will be used to produce different types of fibres.

T60 See RT60.

Tearing strength The resistance of a material to a force tending to tear it apart, measured as the maximum tension the material can withstand without tearing.

Tenacity Tenacity is the customary measure of strength of a fibre or yarn. It is usually defined as the ultimate (breaking) force of the fibre (in gram-force units) divided by the denier. As denier is a measure of the linear density, the tenacity works out to be not a measure of force per unit area, rather a quasi-dimensionless measure analogous to specific strength.

Tensile property Tensile property is the fundamental property of a material. This is a simple test in which a sample is subjected to a controlled tension until failure. The results from the test are commonly used to select a material for an application, for quality control, and to predict how a material will react under other types of forces.

Thermoplastic material A thermoplastic material is a polymeric material that becomes pliable or mouldable above a specific temperature and solidifies upon cooling. Most thermoplastics have a high molecular weight. The polymer chains associate through intermolecular forces, which weaken rapidly with increased temperature, yielding a viscous liquid. Thus, thermoplastics may be reshaped by heating and are typically used to produce parts by various polymer processing techniques such as injection moulding.

Thermosetting material Thermoset or thermosetting, plastics are synthetic materials that strengthen during being heated, but cannot be successfully remoulded or reheated after their initial heat-forming. This is in contrast to thermoplastics, which soften when heated and harden and strengthen after cooling.

Three dimensional (3D) orthogonal woven fabrics 3D woven fabrics are produced on a special 3D loom. The architecture of the 3D orthogonal woven fabric consists of three sets of yarns; warp yarns (y-yarn), weft yarns (x-yarn) and orthogonal-yarns (z-yarn). z-yarn is placed in through-thickness direction of the fabric. In 3D orthogonal woven fabric there is no interlacing between warp and weft yarns and they are straight and perpendicular to each other. On the other hand, z-yarns combine the warp and the weft layers by interlacing (moving up and down) along the y-direction over the weft yarn. Interlacing occurs on the top and the bottom surface of the fabric.

Threshold of hearing The lowest level sound that can be perceived by the human auditory system. This is close to the standard reference level of sound pressure, 20 μPa.

Tone A musical or vocal sound with reference to its pitch, quality, and strength.

Tortuosity The concept of tortuosity is used to characterize the structure of a porous media, to estimate their electrical and hydraulic conductivity. Tortuosity is used to quantify the inertial coupling between the fluid and solid phases of a porous material. In other words, tortuosity is the length of the liquid or gas going through the textile material. Tortuosity affects the sound insulation properties of the porous media. Tortuosity is a parameter without any unit as it is the ratio of two similar units.

Transfer function The output to input relationship of a structure. Mathematically it is the Laplace transform of the output divided by the Laplace transform of the input.

Transfer matrix method It is a method used in optics and acoustics to analyze the propagation of electromagnetic or acoustic waves through a stratified (layered) medium.

Transform A transform is a mathematical operation that converts a function from one domain to another domain with no loss of information. For example, the Fourier transform converts a function of time into a function of frequency.

Transverse wave A transverse wave is a wave where the oscillations occur perpendicular to the direction of energy transfer. If a transverse wave is moving in the positive x-direction, its oscillations are in up and down directions that lie in the y–z plane.

Tuck spacer fabric This fabric consists of a top and a bottom plain knitted layer. These two layers are interconnected with a mesh of yarn. This structure thus represents a tight mesh or bundle of yarn sandwiched between two plain knitted layers. These fabrics are light in weight, flexible and can be knitted with any design to suit the original equipment manufacturer (OEM) and automotive brand requirements.

Tufting Tufting is an ancient technique for making warm garments, especially mittens.

Ultrasound Ultrasound is the sound wave with the frequency higher than the upper audible limit of human hearing (20 kHz). Ultrasound is not different from normal sound in its physical properties; however, humans cannot hear it.

Urethane See Polyurethane.

Polyurethane polymers are traditionally and most commonly formed by reacting a di- or polyisocyanate with a polyol. Both the isocyanates and polyols used to make polyurethanes contain on average two or more functional groups per molecule.

Vacuum-Assisted Resin Transfer Moulding Vacuum-Assisted Resin Transfer Moulding (VARTM) or Vacuum-Injected Moulding (VIM) is a closed mould, out of autoclave (OOA) composite manufacturing process. VARTM is a variation of Resin Transfer Moulding (RTM) with its distinguishing characteristic being the replacement of the top portion of a mould tool with a vacuum bag and the use of a vacuum to assist in resin flow [2]. The process involves the use of a vacuum to facilitate resin flow into a fibre layup contained within a mould tool covered by a vacuum bag. After the impregnation occurs the composite part is allowed to cure at room temperature with an optional post cure sometimes carried out.

Viscoelasticity Viscoelasticity is the property of materials that exhibit both viscous and elastic characteristics when undergoing deformation. Viscous materials such as honey, resist shear flow and strain linearly with time when a stress is applied.

Volume Colloquial equivalent of sound level.

Volumetric mass density The density, or more precisely, the volumetric mass density, of a substance is its mass per unit volume. The symbol most often used for density is ρ (the lower case Greek letter rho).

Vulcanization Vulcanization is a chemical process for converting natural rubber or related polymers into more durable materials via the addition of sulphur or other equivalent curatives or accelerators. These additives modify the polymer by forming cross-links (bridges) between individual polymer chains.

Watt (W) The unit of electrical or acoustical power.

Wattage It is the unit of power used to rate the output of audio amplifiers. For wattage number to have meaning the distortion level and impedance must also be specified.

Wave In Physics, a wave is an oscillation accompanied by a transfer of energy that travels through a medium. Waves are characterized by wavelength, frequency, and the speed at which they move.

Wavelength The wavelength (λ) of a wave is the distance from any point on one wave to the same point on the next wave along (see Amplitude for diagram). To avoid confusion, it can be measured from the top of a crest to the top of the next crest, or from the bottom of a trough to the bottom of the next trough. Wavelength is measured in metres (m) or centimetres (cm). The wavelength of sound at any frequency can be obtained by dividing the speed of sound by the frequency.

Woven fabric Woven fabric is a textile material produced by weaving. It is produced on a loom and made of many threads woven on a warp and a weft. The fabric's integrity is maintained by the mechanical interlocking of the yarns.

Young's modulus It is also known as the elastic modulus. It is the mechanical property of a linear elastic material, which defines the relationship between stress (force per unit area) and strain (proportional deformation) of the material. Young's modulus was named after the British scientist Thomas Young in nineteenth century.

Index

Note: Page numbers followed by *f* and *t* indicate figures and tables respectively

R. Padhye and R. Nayak (eds.), *Acoustic Textiles*, Textile Science and Clothing Technology, DOI 10.1007/978-981-10-1476-5

Printed in the United States
By Bookmasters